Lecture Notes in Mathematics

A collection of informal reports and seminars
Edited by A. Dold, Heidelberg and B. Eckmann, Zürich

62

Harish-Chandra
Institute for Advanced Study, Princeton, New Jersey

Automorphic Forms on Semisimple Lie Groups

Notes by J. G. M. Mars
Mathematisches Institut der Universität Utrecht

1968

Springer-Verlag Berlin · Heidelberg · New York

Library of Congress Catalog Card Number 68-55417. Printed in Germany. Title No. 3668

Preface

These lectures are largely based on an important but, unfortunately, as yet unpublished, manuscript of R. P. Langlands, with the title "On the functional equations satisfied by Eisenstein Series." However they do not cover the last and the most difficult part (Chapter VII) of this manuscript. Langlands himself has given a brief account of his work in a short paper ("Eisenstein Series", pp. 235-252, Algebraic Groups and Discontinuous Subgroups, 1966, Amer. Math. Soc.).

In the last ten years, the analytic theory of automorphic forms has been pushed forward mainly through the contributions of Selberg, Gelfand and Piatetsky-Shapiro, and Langlands. In my opinion, here Selberg's ideas were decisive. But since only a very sketchy account of Selberg's work is available (see "Discontinuous groups and harmonic analysis", pp. 177-189, Proceedings of the International Congress of Mathematicians, 1962), they have not attracted the notice which they undoubtedly deserve. After some of the arithmetic aspects of the theory of discontinuous groups began to be understood, it became clear that one needed analytical tools, in order to apply this newly acquired knowledge to the theory of automorphic forms. It is here that Langlands succeeded in adapting and extending Selberg's methods so as to fit them to the more general situation.

Let me now give an outline of the main steps which lead to the analytic continuation and the functional equations of the Eisenstein series.

1) Definition of the space of cusp forms and proof of the theorem of Gelfand and Piatetsky-Shapiro which says that for any $\alpha \in C_c^{\infty}(G)$, the operator ${}^{0}\lambda(\alpha)$ is compact (Theorem 2, §2, Chap. I). We also verify that $\dim \mathcal{A}(G/\Gamma,\sigma,\chi) < \infty$ (Theorem 1, §2, Chap. I) and prove a simple but important result of Langlands (Theorem 4, §5, Chap. I).

2) Definition and elementary properties of Eisenstein series corresponding to a cuspidal subgroup P of rank q. In particular we show that this series lies in $\mathcal{A}_q(G/\Gamma,\sigma)$ (see Chap. II, §§4,8). Also we derive a certain scalar product formula (Lemma 40).

3) Given $f \in \mathcal{A}_q(G/\Gamma,\sigma)$, we obtain an estimate for $|f|$ in terms of its constant terms f_P corresponding to the cuspidal subgroups P of rank q (Theorems 6 and 6' of Chap. III).

4) The Maass-Selberg relations for forms of type 1. Previously Selberg had considered these relations only in the case $l = 1$ ($l = \operatorname{rank}_{\mathbb{Q}} G$). But Langlands found, for arbitrary l, a dissection of G/Γ into disjoint sets, which imitates Selberg's method of cutting off the cusps. By using this dissection, it became possible to extend Selberg's ideas to the case $l > 1$ (Chap. IV, §§1, 2).

5) Proof of the analytic continuation and the functional equation of the Eisenstein series of type 1. Here we follow the method of Selberg (Chap. IV, §§5-8).

6) An induction on $q = \operatorname{rank} P$, which enables us to obtain similar results for Eisenstein series of type q (Theorems 8 and 9, Chap. V, §2). This is entirely due to Langlands.

I am very grateful to J. G. M. Mars for having taken the trouble of preparing these notes from my lectures, which were given at the Institute for Advanced Study, Princeton, during '66-'67. Apart from a careful attention to details, this has required a certain reorganization of the material and although I have not personally checked all the proofs, I trust that they do not contain any substantial mistakes.

Harish-Chandra

Introduction

Let G be a connected semisimple Lie group with finite center and Γ a closed subgroup of G, which we assume to be unimodular. Let μ denote the invariant measure on G/Γ. We have the obvious representation λ of G on $L_2(G/\Gamma) = L_2(G/\Gamma,\mu)$. The main problem here is to carry out, as explicitly as possible, the reduction of λ. Let K be a maximal compact subgroup of G. In case $\Gamma = K$, this problem has been solved. Similarly if $\Gamma = \{1\}$, it has been more or less solved. (It is then essentially the same problem as the explicit determination of the Plancherel formula.)

Let $\mathcal{E}$ be the set of all equivalence classes of irreducible unitary representations of G. Fix $\omega \in \mathcal{E}$. We say that an element $f \in L_2(G/\Gamma)$ is of type ω, if the smallest closed invariant subspace V_f of $L_2(G/\Gamma)$ containing f, is irreducible under λ and the corresponding representation λ_f lies in ω. Let $\mathcal{H}_\omega$ be the smallest closed subspace of $L_2(G/\Gamma)$ containing all elements of type ω. Then $\mathcal{H}_\omega$ can be written as an orthogonal sum of irreducible subspaces. Let $m(\omega)$ denote the number of irreducible subspaces in this sum. Then $m(\omega)$ is called the multiplicity of ω in λ. It is an important problem to compute $m(\omega)$.

ω is said to occur discretely in λ, if $m(\omega) > 0$. For example, if G is not compact and $\Gamma = K$, $m(\omega) = 0$ for all ω and therefore λ has no discrete components. If $\Gamma = \{1\}$, the discrete components occur if and only if G has a compact Cartan subgroup. Moreover one then knows which discrete components occur. Finally $m(\omega) = \infty$ for every discrete component ω, provided G is not compact.

A class $\omega \in \mathcal{E}$ is said to be L_p $(p = 1, 2)$ if some matrix coefficient of ω lies in $L_p(G)$. If Γ is a discrete subgroup of G such that G/Γ is compact, then $m(\omega) < \infty$ for every ω and $L_2(G/\Gamma) = \mathrm{Cl}(\sum_\omega \mathcal{H}_\omega)$. Moreover if ω is an L_1-class, there exists an explicit formula for $m(\omega)$. It is important to generalize this formula in two directions. Firstly we should drop the assumption that ω is L_1 and, secondly, we should relax the condition that G/Γ be compact. However no general results in this direction have been obtained so far.

From now on we assume that Γ is a discrete subgroup of G such that G/Γ has finite invariant measure. For $G = SL(2,\mathbb{R})$ and $\Gamma = SL(2,\mathbb{Z})$ (and perhaps also for some other Γ), the problem of the reduction of λ was solved by Selberg. The case for all $\chi \in X_Q(Z(\underline{\underline{A}}))$. Put $\underline{\underline{M}} = \bigcap_{\chi} \mathrm{Ker}(\chi^2)$, where χ runs through $X_Q(Z(\underline{\underline{A}}))$. Then $\underline{\underline{M}}$ is a Q-subgroup of $Z(\underline{\underline{A}})$ and we have $Z(\underline{\underline{A}}) = \underline{\underline{M}}\,\underline{\underline{A}}$, $\underline{\underline{P}} = \underline{\underline{M}}\,\underline{\underline{A}}\,\underline{\underline{U}}$. $\underline{\underline{M}} \cap \underline{\underline{A}}$ is finite.

We call $\underline{\underline{A}}$ a split component of $\underline{\underline{P}}$. Any two split components of $\underline{\underline{P}}$ are conjugate by a unique element of $\underline{\underline{U}}_Q$. When $\underline{\underline{A}}$ has been chosen, $\underline{\underline{M}}$ is of course determined.

It is easy to see that $\underline{\underline{M}}$ satisfies the same conditions as $\underline{\underline{G}}$. If 2'), resp. 2") holds for $\underline{\underline{G}}$, then 2'), resp. 2") holds for $\underline{\underline{M}}$.

Now we consider the points over R of the groups we have introduced and put $G = \underline{\underline{G}}_R$, $P = \underline{\underline{P}}_R$, $U = \underline{\underline{U}}_R$, A = connected component of $\underline{\underline{A}}_R$, $M = \underline{\underline{M}}_R$.
We have $P = (\underline{\underline{M}}\,\underline{\underline{A}})_R U$ and one proves as above for the complex groups that $(\underline{\underline{M}}\,\underline{\underline{A}})_R = MA$, hence $P = MAU$. Obviously $M \cap A = \{1\}$, so the decomposition of an element p of P in the form $p = mau$ $(m \in M,\ a \in A,\ u \in U)$ is unique. We call P a <u>cuspidal subgroup</u> of G, A a <u>split component</u> of P, (P,A) a <u>cuspidal pair</u> of G and the decomposition $P = MAU$ a <u>Langlands decomposition</u> of P.

Let $\mathfrak{g}$ be the Lie algebra of G; $\mathfrak{g}$ is the set of points over R of the Lie algebra $\mathfrak{g}_c$ of (the connected component of) $\underline{\underline{G}}$. Also let $\mathfrak{p}, \mathfrak{n}, \mathfrak{a}, \mathfrak{m}$ denote the Lie algebras of P, U, A, M respectively. Then $\mathfrak{p} = \mathfrak{m} + \mathfrak{a} + \mathfrak{n}$; $\mathfrak{m} + \mathfrak{a}$ is equal to the centralizer $\mathfrak{z}(\mathfrak{a})$ of $\mathfrak{a}$ in $\mathfrak{g}$; $\mathfrak{m}$ is the orthogonal complement of $\mathfrak{a}$ in $\mathfrak{z}(\mathfrak{a})$ (with respect to the Killing form). We call $(\mathfrak{p}, \mathfrak{a})$ a <u>cuspidal pair</u> of $\mathfrak{g}$.

Given $\lambda \in \mathfrak{a}^*$ (= dual of the real vectorspace $\mathfrak{a}$), let $\mathfrak{n}_\lambda$ denote the set of all $X \in \mathfrak{n}$ such that $[H,X] = \lambda(H)X$ for $H \in \mathfrak{a}$. An element λ of $\mathfrak{a}^*$ is called a <u>root</u> of $(\mathfrak{p}, \mathfrak{a})$ if $\mathfrak{n}_\lambda \neq 0$. The set of all roots of $(\mathfrak{p}, \mathfrak{a})$ is denoted by $\Sigma(\mathfrak{p}\,|\,\mathfrak{a})$, or $\Sigma(P|A)$, or Σ. It is known that $0 \notin \Sigma$ and that, if $l = \dim_R \mathfrak{a}$, there exist l linearly independent roots $\alpha_1, \ldots, \alpha_l$ such that every root $\alpha \in \Sigma$ can be written in the form $\alpha = m_1\alpha_1 + \ldots + m_l\alpha_l$ with $m_i \in Z$, $m_i \geq 0$. The set $\{\alpha_1, \ldots, \alpha_l\}$ is the set of <u>simple roots</u> of $(\mathfrak{p}, \mathfrak{a})$ and is denoted by $\Sigma^o(\mathfrak{p}\,|\,\mathfrak{a}) = \Sigma^o(P|A) = \Sigma^o$.

Let $(\mathfrak{p}, \mathfrak{a})$ be a cuspidal pair of $\mathfrak{g}$ and F any subset of $\Sigma^o(\mathfrak{p}\,|\,\mathfrak{a})$. Let $\mathfrak{a}_F$ denote the subspace of $\mathfrak{a}$ consisting of the elements H of $\mathfrak{a}$ such that $\alpha(H) = 0$ for all $\alpha \in F$. Put $\mathfrak{n}_F = \sum_{\alpha \in \Sigma_F'} \mathfrak{n}_\alpha$, where Σ_F' is the set of all elements $\alpha \in \Sigma$ which do not vanish identically on $\mathfrak{a}_F$, and define $\mathfrak{p}_F = \mathfrak{z}(\mathfrak{a}_F) + \mathfrak{n}_F$.

Finally, let P_F be the normalizer of $\mathfrak{p}_F$ in G and U_F, resp. A_F, the analytic subgroup of G corresponding to $\mathfrak{n}_F$, resp. $\mathfrak{a}_F$. Then (P_F, A_F) is a cuspidal pair of G and U_F is the unipotent radical of P_F. If (P,A) is the cuspidal pair of G corresponding to $(\mathfrak{p}, \mathfrak{a})$ and U is the unipotent radical of P, it is clear that $P_F \supset P$, $A_F \subset A$, $U_F \subset U$. The roots of $(\mathfrak{p}_F, \mathfrak{a}_F)$ are the restrictions to $\mathfrak{a}_F$ of the elements of Σ_F', the simple roots are the restrictions of the elements of ${}^cF$ (${}^cF$ denotes the complement of F in $\Sigma^o(P|A)$).

<u>Definition</u>. Let (P,A) and (P',A') be two cuspidal pairs of G. We say that (P,A) dominates (P',A') and write $(P,A) \succ (P',A')$, if $P \supset P'$, $A \subset A'$.

So in the above situation (P_F, A_F) dominates (P,A). Moreover, for any cuspidal group P' containing P, there exists one and only one subset F of $\Sigma^o(P|A)$ such that $P' = P_F$; this follows immediately from the same assertion for the connected group $\underline{\underline{G}}^o$ which is well known.

Let $\underline{\underline{P}}$ be any parabolic subgroup of $\underline{\underline{G}}$ and $\underline{\underline{A}}$ a split component of $\underline{\underline{P}}$. We put ${}^o\underline{\underline{P}} = \cap \operatorname{Ker}(\chi^2)$, where χ runs through $X_Q(\underline{\underline{P}})$. Clearly ${}^o\underline{\underline{P}} = \underline{\underline{M}}\,\underline{\underline{U}}$. We define ${}^oP = ({}^o\underline{\underline{P}})_R$. Then ${}^oP = MU$, and ${}^oP/U = MU/U = M$; this gives a canonical homomorphism $\pi_{P|M} : {}^oP \to M$.

Let (P,A) be a cuspidal pair of G and $P = MAU$ the Langlands decomposition. Let $({}^*P, {}^*A)$ be a cuspidal pair of M and ${}^*P = {}^*M{}^*A{}^*U$ the Langlands decompostion. Put $M' = {}^*M$, $A' = {}^*AA$, $U' = {}^*UU$, $P' = M'A'U'$. Then (P',A') is a cuspidal pair of G with the Langlands decomposition $P' = M'A'U'$; we have $P' \subset P$.

<u>Lemma 2</u>. The correspondence ${}^*P \longmapsto P'$ gives a bijection of the set of cuspidal subgroups of M on the set of those cuspidal subgroups of G which are contained in P. The inverse map is given by $Q \longmapsto \pi_{P|M}(Q \cap {}^oP)$.

<u>Proof</u>. It is obvious that, in the above notation, we have ${}^*P = \pi_{P|M}(P' \cap {}^oP)$. One sees easily that, if Q is a cuspidal subgroup of G contained in P, $\pi_{P|M}(Q \cap {}^oP)$ is a cuspidal subgroup of M; call it ${}^*P$. Then the corresponding P' is equal to ${}^*PAU = (Q \cap M)AU = Q$.

<u>Definition</u>. We call P <u>mincuspidal</u> if it is minimal among cuspidal subgroups of G. A cuspidal pair (P,A) is called mincuspidal if P is mincuspidal.

$$c(st:\lambda) = c(s:t\lambda)c(t:\lambda)$$

and

$$E_\lambda = c(s:\lambda)E_{s\lambda} \qquad (s,t \in W).$$

Since E_λ is actually defined by means of a Dirichlet series, there is a certain obvious analogy with the ζ-functions and L-series of number theory. In particular the functions $c(s:\lambda)$ seem to have a product formula. On the other hand there is also a strong analogy with the theory of elementary spherical functions. If we put

$$\phi_\lambda(x) = \int_K e^{(\lambda-\varrho)(H(xk))}dk \qquad (x \in G),$$

then it is known that $\phi_{s\lambda} = \phi_\lambda$ $(s \in W)$.

For $G = SL(n,R)$ and $\Gamma = SL(n,Z)$, Selberg had obtained the analytic continuation of all the Eisenstein series (and not just those mentioned above). Langlands has now done this in the general case. Actually he does not confine himself to the arithmetic case but makes a certain set of assumptions on (G,Γ). In view of the recent work of two Russians, Vinberg and Makarov, where they construct non-arithmetic discrete groups Γ such that G/Γ has finite measure, it seems conceivable that the assumptions of Langlands are more general. However they are rather unwieldly and so we shall confine ourselves to the arithmetic case.

Contents

Chapter V

CHAPTER I

§1. Algebraic and arithmetic results.

The name "algebraic group" will be given here exclusively to algebraic subgroups of $GL(n,C)$ (any n). A $\mathbb{Q}$-group is an algebraic group which is defined over $\mathbb{Q}$. Expressions like $\mathbb{Q}$-subgroup are used in the obvious sense. If $\underline{\underline{G}}$ is a connected algebraic group, we denote by $\mathfrak{g}_C$ the Lie algebra of $\underline{\underline{G}}$, that is, the Lie algebra of all left-invariant C-linear derivations of the field $\mathfrak{F}(\underline{\underline{G}})$ of rational functions on $\underline{\underline{G}}$. If $\underline{\underline{G}}$ is defined over $\mathbb{Q}$ we define $\mathfrak{g}_Q$ as the Lie algebra of all left-invariant $\mathbb{Q}$-linear derivations of the field $\mathfrak{F}_Q(\underline{\underline{G}})$ of rational functions defined over $\mathbb{Q}$ on $\underline{\underline{G}}$. Then $\mathfrak{g}_C = C \otimes_{\mathbb{Q}} \mathfrak{g}_{\mathbb{Q}}$ and $\mathfrak{g}_Q$ is the set of points over $\mathbb{Q}$ of the algebraic variety $\mathfrak{g}_C$; a subspace $\mathfrak{n}$ of $\mathfrak{g}_C$ is a $\mathbb{Q}$-subspace, i.e. is defined over $\mathbb{Q}$, if and only if $\mathfrak{n} = C \otimes_{\mathbb{Q}} \mathfrak{n}_Q$, where $\mathfrak{n}_Q = \mathfrak{n} \cap \mathfrak{g}_Q$. If $\underline{\underline{H}}$ is a connected algebraic subgroup of the $\mathbb{Q}$-group $\underline{\underline{G}}$, then $\underline{\underline{H}}$ is defined over $\mathbb{Q}$ if and only if its Lie algebra $\mathfrak{h}_C$ is a $\mathbb{Q}$-subspace of $\mathfrak{g}_C$.

For any algebraic group $\underline{\underline{G}}$, $X(\underline{\underline{G}})$ will denote the group of rational characters of $\underline{\underline{G}}$ (rational homomorphisms $\underline{\underline{G}} \to C^*$), and if $\underline{\underline{G}}$ is defined over $\mathbb{Q}$, $X_Q(\underline{\underline{G}})$ is the subgroup of $X(\underline{\underline{G}})$ consisting of those characters which are defined over $\mathbb{Q}$. The connected component of a group $\underline{\underline{G}}$ will be denoted by $\underline{\underline{G}}^o$.

An algebraic group $\underline{\underline{G}}$ is called reductive when the radical of its connected component is a torus. The Lie algebra of a reductive group is reductive, i.e. is direct sum of a semisimple Lie algebra and an abelian one.

We assume from now on that we are given a reductive $\mathbb{Q}$-group $\underline{\underline{G}}$ satisfying the following conditions:

1) for any $\chi \in X_Q(\underline{\underline{G}})$, we have $\chi^2 = 1$;
2) if $\underline{\underline{T}}$ is a maximal $\mathbb{Q}$-split torus in $\underline{\underline{G}}$, then $Z(\underline{\underline{T}})$ (centralizer of $\underline{\underline{T}}$ in $\underline{\underline{G}}$) meets every connected component of $\underline{\underline{G}}$.

[The condition 1) implies that $X_Q(\underline{\underline{G}}^o) = \{1\}$. For various reasons it is necessary to have this. It is for technical reasons, connected with the reduction to anisotropic kernels, that we assume 1) and 2)].

Consider also the conditions:

2') If $\underline{\underline{T}}$ is a maximal torus of $\underline{\underline{G}}$, then $Z(\underline{\underline{T}})$ meets every connected component of $\underline{\underline{G}}$.

2") The center $\underline{\underline{Z}}$ of $\underline{\underline{G}}$ meets every connected component of $\underline{\underline{G}}$.

Obviously 2") implies 2'), and 2') implies 2). For reasons to be explained later, we shall assume from §2 on that $\underline{\underline{G}}$ satisfies the condition 2').

By a parabolic subgroup of a connected $\mathbb{Q}$-group $\underline{\underline{H}}$ we mean a $\mathbb{Q}$-subgroup of $\underline{\underline{H}}$ which contains a Borel subgroup of $\underline{\underline{H}}$. Such a subgroup is always connected and equal to its normalizer in $\underline{\underline{H}}$.

Definition. We call parabolic subgroup of $\underline{\underline{G}}$ the normalizer in $\underline{\underline{G}}$ of any parabolic subgroup of $\underline{\underline{G}}^{o}$.

Lemma 1. If $\underline{\underline{P}}$ is a parabolic subgroup of $\underline{\underline{G}}$, $\underline{\underline{P}}^{o}$ is a parabolic subgroup of $\underline{\underline{G}}^{o}$, $\underline{\underline{P}}$ is the normalizer in $\underline{\underline{G}}$ of $\underline{\underline{P}}^{o}$, $\underline{\underline{P}} \cap \underline{\underline{G}}^{o} = \underline{\underline{P}}^{o}$ and $\underline{\underline{P}}$ meets every connected component of $\underline{\underline{G}}$.

Let $\underline{\underline{P}}$ be a parabolic subgroup of $\underline{\underline{G}}$. By definition there exists a parabolic subgroup $\underline{\underline{Q}}$ of $\underline{\underline{G}}^{o}$ such that $\underline{\underline{P}} = N(\underline{\underline{Q}})$ (= normalizer of $\underline{\underline{Q}}$ in $\underline{\underline{G}}$). Then $\underline{\underline{P}} \cap \underline{\underline{G}}^{o}$ is the normalizer of $\underline{\underline{Q}}$ in $\underline{\underline{G}}^{o}$ and therefore $P \cap \underline{\underline{G}}^{o} = \underline{\underline{Q}} = \underline{\underline{P}}^{o}$. Let $\underline{\underline{U}}$ be the unipotent radical of $\underline{\underline{P}}^{o}$. $\underline{\underline{P}}^{o}$ has a Levi decomposition $\underline{\underline{P}}^{o} = Z^{o}(\underline{\underline{A}})\underline{\underline{U}}$, where $\underline{\underline{A}}$ is a $\mathbb{Q}$-split torus and $Z^{o}(\underline{\underline{A}})$ is the centralizer of $\underline{\underline{A}}$ in $\underline{\underline{G}}^{o}$ ($Z^{o}(\underline{\underline{A}})$ is connected and therefore equal to the connected component of the centralizer $Z(\underline{\underline{A}})$ of $\underline{\underline{A}}$ in $\underline{\underline{G}}$). $Z^{o}(\underline{\underline{A}})$ is reductive and $\underline{\underline{A}}$ is the maximal $\mathbb{Q}$-split torus of the radical of $Z^{o}(\underline{\underline{A}})$. Since the conjugation by an element of $Z(\underline{\underline{A}})$ induces an automorphism of $\underline{\underline{G}}^{o}$ which leaves $\underline{\underline{A}}$ pointwise fixed, so leaves $\underline{\underline{U}}$ fixed and $Z^{o}(\underline{\underline{A}})$ fixed, so leaves $\underline{\underline{P}}^{o}$ fixed, we have $Z(\underline{\underline{A}}) \subset N(\underline{\underline{P}}^{o}) = \underline{\underline{P}}$. On the other hand, the hypothesis 2) implies that $Z(\underline{\underline{A}})$ meets every connected component of $\underline{\underline{G}}$. Since $\underline{\underline{P}} \cap \underline{\underline{G}}^{o}$ is connected, the intersection of $\underline{\underline{P}}$ with a connected component of $\underline{\underline{G}}$ consists of one connected component of $\underline{\underline{P}}$; so $Z(\underline{\underline{A}})\underline{\underline{U}}$ meets every connected component of $\underline{\underline{P}}$ and contains $\underline{\underline{P}}^{o}$, hence $\underline{\underline{P}} = Z(\underline{\underline{A}})\underline{\underline{U}}$.

The groups $Z(\underline{\underline{A}})/Z^{o}(\underline{\underline{A}})$ and $\underline{\underline{G}}/\underline{\underline{G}}^{o}$ are $\mathbb{Q}$-isomorphic, so they have the same character group. Let χ be a character belonging to $X_{\mathbb{Q}}(Z(\underline{\underline{A}}))$ such that χ^2 vanishes on $\underline{\underline{A}}$. Then χ vanishes on $\underline{\underline{A}}$; it vanishes then also on $Z^{o}(\underline{\underline{A}})$, hence can be regarded as a character of $Z(\underline{\underline{A}})/Z^{o}(\underline{\underline{A}})$, so $\chi^2 = 1$ (from hypothesis 1)). It follows from this, that for any given $x \in Z(\underline{\underline{A}})$, we can find an element a of $\underline{\underline{A}}$ such that $\chi^2(a) = \chi^2(x)$

$G = SL(n,R)$, $\Gamma = SL(n,Z)$ has been considered by Gelfand and Piatetsky-Shapiro. Finally the case when G is any real algebraic semisimple group defined over $\mathbb{Q}$ and Γ any arithmetic subgroup of G, has been studied by Langlands and he has obtained the deepest and the most general results so far. In this case it is possible to show that $m(\omega) < \infty$. However the problem of computing $m(\omega)$ remains largely untouched.

There is another way of approaching the problem of the reduction of $L_2(G/\Gamma)$. Let Let $\mathfrak{Z}$ be the algebra of all differential operators on G which commute with both left and right translations. Let $\mathfrak{g}$ be the Lie algebra of G and $\mathfrak{G}$ the universal enveloping algebra of $\mathfrak{g}_c$. Then $\mathfrak{G}$ may be identified with the algebra of all left-invariant differential operators on G and $\mathfrak{Z}$ is the center of $\mathfrak{G}$. Put $\mathcal{H} = L_2(G/\Gamma)$ and let $\mathcal{H}_\infty$ denote the subspace of all $f \in \mathcal{H}$ such that the mapping $x \to \lambda(x)f$ of G into $\mathcal{H}$ is of class C^∞. Then one gets in a natural way a representation of $\mathfrak{G}$ on $\mathcal{H}_\infty$, which we denote by λ_∞. For any differential operator D on G, let D^* denote its adjoint so that

$$\int Df.gdx = \int f.D^*gdx \qquad (f,g \in C_c^\infty(G)) .$$

(dx is the Haar measure on G.) Then $\mathfrak{G}^* = \mathfrak{G}$ and $\mathfrak{Z}^* = \mathfrak{Z}$. Let η denote the conjugation of $\mathfrak{g}_c$ with respect to $\mathfrak{g}$. We say that an element $z \in \mathfrak{Z}$ is hermitian if $\eta(z^*) = z$.

Lemma. _If_ $z \in \mathfrak{Z}$ _is hermitian, then_ $\lambda_\infty(z)$ _is an essentially self-adjoint operator in the sense of Hilbert-space theory._

Now $\mathcal{H}_\infty$ is stable under $\lambda(x)$ $(x \in G)$ and $\lambda_\infty(z)$ $(z \in \mathfrak{Z})$ commutes with $\lambda(x)$. Hence we may regard our reduction problem, at least as a first approximation, as that of obtaining a simultaneous spectral decomposition for all the operators $\lambda_\infty(z)$ $(z \in \mathfrak{Z})$. So, roughly speaking, it becomes an eigenfunction expansion problem. Given $f \in L_2(G/\Gamma)$, we have to express it as a linear combination of eigenfunctions of $\mathfrak{Z}$ on G/Γ. Fix a maximal compact subgroup K of G. There is no essential loss of generality in assuming that f is K-finite on the left. Then the eigenfunctions that we are looking for, may also be assumed to be left K-finite. It is not difficult to show that such eigenfunctions are in fact analytic.

Thus it is natural to introduce the following definition. Let σ be a unitary representation of K on a finite-dimensional complex Hilbert space V and χ a character of $\mathfrak{Z}$ i.e. a homomorphism of $\mathfrak{Z}$ into C such that $\chi(1) = 1$. By an automorphic form of type (σ,χ), we mean a C^∞ function $f : G/\Gamma \to V$ such that $f(kx) = \sigma(k)f(x)$ $(k \in K, x \in G)$ and $zf = \chi(z)f$ $(z \in \mathfrak{Z})$. However one finds that some of these eigenfunctions have nothing to do with harmonic analysis on G/Γ, since they grow much too fast at infinity. Therefore we impose, in addition, a mild growth condition on f, so as to exclude these extraneous functions.

Let $\mathcal{A}(G/\Gamma,\sigma,\chi)$ denote the space of all automorphic forms of type (σ,χ). Then the first result is that $\dim \mathcal{A}(G/\Gamma,\sigma,\chi) < \infty$ (provided Γ satisfies certain reasonable conditions). For example this is true if $G = SL(n,R)$, $\Gamma = SL(n,Z)$. In the case of $G = Sp(n,R)$, $\Gamma = Sp(n,Z)$ and holomorphic forms, this result was first proved by Siegel.

Now a few words about how to construct such eigenfunctions. When G/Γ is compact, no general method of constructing them explicitly is known (except when G has a discrete series). However their existence is assured from the fact that $L_2(G/\Gamma)$ reduces, in this case, into a discrete orthogonal sum of irreducible subspaces. Every such subspace then consists of eigenfunctions.

So let us now consider the case when G/Γ is not compact e.g. $G = SL(n,R)$, $\Gamma = SL(n,Z)$. Then we have the Iwasawa decomposition $G = KAN$, where $K = SO(n)$, A = the group of diagonal matrices in G with all diagonal elements > 0, and $N = \left\{\begin{pmatrix}1 & * \\ 0 & 1\end{pmatrix}\right\}$. Let $x = kan$ $(k \in K,\ a \in A,\ n \in N)$. We write $k(x) = k$ and $H(x) = \log a \in \mathfrak{a}$ = Lie algebra of A. Then

$$dx = e^{2\varrho(\log a)}\,dk\,da\,dn$$

where ϱ is a certain linear function on $\mathfrak{a}$. Let σ be a unitary representation of K on a finite-dimensional space V as above. Then for any linear function λ on $\mathfrak{a}_c$, the function

$$x \longrightarrow \sigma(k(x))e^{(\lambda-\varrho)(H(x))} \qquad (x \in G)$$

from G to the space $\mathcal{E}(V)$ of endomorphisms of V, is an eigenfunction of $\mathfrak{Z}$. Let χ_λ denote the corresponding character of $\mathfrak{Z}$. Then $\chi_{s\lambda} = \chi_\lambda$ for $s \in W$ (= Weyl group of $(\mathfrak{g},\mathfrak{a})$). This function may be considered as a function on G/N. Therefore if the series

$$E_\lambda(x) = \sum_{\gamma \in \Gamma/\Gamma \cap N} \sigma(k(x\gamma))e^{(\lambda-\varrho)(H(x\gamma))}$$

converges, it gives an eigenfunction of $\mathfrak{Z}$ on G/Γ. Now the series does converge for suitable λ. But the corresponding χ_λ is such that it is clearly not the infinitesimal character of an irreducible unitary representation of G. Therefore we are forced to study $E_\lambda(x)$ as a function of λ and show that, by analytic continuation, it can be extended to a meromorphic function of λ on $\mathfrak{a}_c^*$ (= dual of $\mathfrak{a}_c$).

For simplicity, let us consider the case when $\sigma = 1$ and $V = C$. Normalize the Haar measure dn on N in such a way that $N/N \cap \Gamma$ has measure 1. Then

$$e^{\varrho(\log a)} \int_{N/N \cap \Gamma} E_\lambda(an)\,dn = \sum_{s \in W} c(s:\lambda)e^{s\lambda(\log a)} \qquad (a \in A),$$

where $c(s:\lambda)$ are meromorphic functions on $\mathfrak{a}_c^*$ and $c(1:\lambda) = 1$. Moreover we have the functional equations

Definition. The rank of a cuspidal group P is the dimension of any split component of P.

The mincuspidal subgroups of G are the cuspidal subgroups of G with maximal rank, i.e. with rank equal to $\mathrm{rank}_{\mathbb{Q}}(G)$.

We have the following known result.

Let (P_o, A_o) be a mincuspidal pair of G and $\Sigma^o = \Sigma^o(P_o|A_o)$. Then, given any cuspidal pair (P,A) of G, there exists an element ξ of $G_{\mathbb{Q}}$ and a unique subset F of Σ^o such that ${}^{\xi}(P,A) = ((P_o)_F, (A_o)_F)$.

[If $x \in G$ and $S \subset G$, ${}^{x}S$ means xSx^{-1}]

The cuspidal subgroups of rank 1 are called maximal cuspidal subgroups. They are maximal among the cuspidal subgroups $\neq G$ and exist only if $\mathrm{rank}_{\mathbb{Q}}(G) > 0$.

Now we fix once for all a maximal compact subgroup K of G. K meets every connected component of G. For any cuspidal subgroup P of G we have $G = KP$. If (P,A) is a cuspidal pair of G, $P = MAU$, then $K \cap P = K \cap {}^{o}P$. It follows from this, that in a decomposition $x = k\,mau$ $(k \in K, m \in M, a \in A, u \in U)$ of an element x of G, a is unique; we denote it sometimes by $a(x)$; $\log a$ will often be denoted by $H(x)$ or $H_{P|A}(x)$.

In the same notation, we put $K_M = \pi_{P|M}(K \cap {}^{o}P)$. K_M is a maximal compact subgroup of M and $\pi_{P|M}$ induces an isomorphism of $K \cap {}^{o}P$ on K_M (since $K \cap U = \{1\}$).

A Siegel domain with respect to a cuspidal pair (P,A) is a subset $\mathfrak{S}$ of G of the form $\mathfrak{S} = KA_t\omega$, where $A_t = \{a \in A : \alpha(\log a) \leq t$ for $\alpha \in \Sigma^o\}$ and ω is a bounded subset of ${}^{o}P$.

An arithmetic subgroup of G is a subgroup Γ of G such that 1) $\Gamma \subset G_{\mathbb{Q}}$, 2) Γ is commensurable with $G_{\mathbb{Z}}$.

The following theorem is due to Borel.

Let (P,A) be a mincuspidal pair and Γ an arithmetic subgroup of G. Then $P_{\mathbb{Q}}\backslash G_{\mathbb{Q}}/\Gamma$ is finite; and if Ξ is a subset of $G_{\mathbb{Q}}$, in order that there exists a Siegel domain $\mathfrak{S}$ with respect to (P,A) such that $G = \mathfrak{S}\,\Xi\,\Gamma$ it is necessary and sufficient that $G_{\mathbb{Q}} = P_{\mathbb{Q}}\Xi\,\Gamma$.

If Γ is an arithmetic subgroup of G and ξ is an element of $G_{\mathbb{Q}}$, then ${}^{\xi}\Gamma$ is an arithmetic subgroup and Γ and ${}^{\xi}\Gamma$ are commensurable. Also, Γ and $\bigcap_{\xi \in \Xi} {}^{\xi}\Gamma$ are commensurable if Ξ is a finite subset of $G_{\mathbb{Q}}$.

For any cuspidal subgroup P of G we have $\Gamma \cap P = \Gamma \cap {}^{o}P$. Put $\Gamma_M = \pi_{P|M}(\Gamma \cap {}^{o}P)$; Γ_M is an arithmetic subgroup of M .

Let (P_o, A_o) be a mincuspidal pair of G . If P is any cuspidal subgroup of G, then P is conjugate under $G_{\mathbb{Q}}$ to one of the groups $(P_o)_F$ $(F \subset \Sigma^o(P_o|A_o))$, so modulo conjugation under $G_{\mathbb{Q}}$ there are but finitely many cuspidal subgroups. Consider the set of cuspidal subgroups which are conjugate under $G_{\mathbb{Q}}$ to a given cuspidal group P . Modulo conjugation under Γ they are in one-to-one correspondence with the elements of $\Gamma \backslash G_{\mathbb{Q}} / P_{\mathbb{Q}}$, which set is finite. Thus we get the result that G has only finitely many cuspidal subgroups modulo conjugation under Γ .

§2. Automorphic forms.

From our hypothesis 1) on $\underline{\underline{G}}$ it follows, since $x \to \det(x)$ is a rational character of $\underline{\underline{G}}$ defined over $\mathbb{Q}$, that $\det(x) = \pm 1$ for every $x \in \underline{\underline{G}}$. If $\underline{\underline{G}} \subset GL(n,\mathbb{C})$ and $x \in \underline{\underline{G}}$ is the matrix (x_{ij}), we define

$$\|x\| = \left(\Sigma |x_{ij}|^2\right)^{1/2} = (\mathrm{tr}(x^* x))^{1/2} .$$

The function $x \to \|x\|$ on $\underline{\underline{G}}$ has the following properties.

1) $\|x\| \geq 1$. This follows from the inequality $(\mathrm{tr}(x^* x))^n \geq |\det(x)|^2$.

2) $\|xy\| \leq \|x\| \, \|y\|$.

3) There are constants c and N such that $\|x^{-1}\| \leq c\|x\|^N$.

We fix now an arithmetic subgroup Γ of G .

If σ is a representation of K on a finite-dimensional vector space V , a function $f : G \to V$ is called a σ-<u>function</u> if $f(kx) = \sigma(k) f(x)$ for all $k \in K$, $x \in G$. We denote by $C^\infty(G,\sigma)$ the space of σ-functions which are of class C^∞ , by $L^2(G/\Gamma,\sigma)$ the space of σ-functions on G which are right-invariant under Γ and are square-integrable on G/Γ , etc.

For any cuspidal pair (P,A) of G, the representation (σ,V) of K permits us to define a representation σ_M of K_M on V :

$$\sigma_M(\pi_{P|M}(k)) = \sigma(k) \qquad (k \in K \cap {}^oP).$$

Definition. Given any finite-dimensional representation (σ,V) of K, we denote by $\mathcal{A}(G/\Gamma,\sigma)$ the space of all functions f in $C(G/\Gamma,\sigma)$ for which there exists a real number r such that $\sup_{x\in G} |f(x)|\,\|x\|^{-r} < \infty$.

Now let $\mathfrak{G}$ be the universal enveloping algebra of $\mathfrak{g}_c$, $\mathfrak{Z}$ the center of $\mathfrak{G}$ and $\mathfrak{Z}^o$ the algebra of invariants of $\underline{G}$ in $\mathfrak{G}$. Then $\mathfrak{Z}^o \subset \mathfrak{Z}$ and $\mathfrak{Z}$ is a finitely generated $\mathfrak{Z}^o$-module. If the condition 2') of §1 is verified by $\underline{G}$, then $\mathfrak{Z}^o = \mathfrak{Z}$. For simplicity we shall suppose that $\underline{G}$ satisfies 2').

The elements of $\mathfrak{G}$ will be regarded as left-invariant differential operators on G; $\mathfrak{Z}$ is then the algebra of bi-invariant differential operators on G. We recall that, if D is a differential operator on G and T a distribution on G, then the distribution DT is defined by $(DT)(f) = T(D^*f)$ $(f \in C_c^\infty(G))$, where D^* is the differential operator defined by

$$\int_G (D^*f)\, g\, dx = \int_G f\,(Dg)\, dx \qquad (f,g \in C_c^\infty(G)).$$

In particular, zf is defined when $z \in \mathfrak{Z}$, $f \in C(G)$.

Definition. Let (σ,V) be a finite-dimensional representation of K. Let χ be a representation of $\mathfrak{Z}$ on V such that the operators $\chi(z)$ $(z \in \mathfrak{Z})$ commute with the operators $\sigma(k)$ $(k \in K)$. Then $\mathcal{A}(G/\Gamma,\sigma,\chi)$ is the space of all $f \in \mathcal{A}(G/\Gamma,\sigma)$ satisfying the condition

$$zf = f\chi(z) \quad \text{for} \quad z \in \mathfrak{Z}.$$

[We use the notation at the right for the operators $\chi(z)$; $f\chi(z)$ is of course the function $x \to f(x)\chi(z)$].

The elements of $\mathcal{A}(G/\Gamma,\sigma,\chi)$ are called <u>automorphic forms</u> of type (σ,χ) (on G with respect to Γ). If f is an automorphic form, the functions zf $(z \in \mathfrak{Z})$, resp. the functions $x \to f(kx)$ $(k \in K)$, form, resp. span, a finite-dimensional vector space; in other words, f is $\mathfrak{Z}$-finite and (left-)K-finite. This implies that f is analytic.

We shall prove the following theorem.

<u>Theorem 1</u>. $\dim \mathcal{A}(G/\Gamma,\sigma,\chi) < \infty$.

Let us now consider the space $L^2(G/\Gamma)$ and the unitary representation λ of G on it: $(\lambda(x)f)(y) = f(x^{-1}y)$.

<u>Lemma 3</u>. For an element f of $L^2(G/\Gamma)$ the following conditions are equivalent.

1) For any cuspidal subgroup P of G different from G and any function $\varphi \in C_c(G/U)$ we have $\int_{G/U\cap\Gamma} \varphi(x)\ f(x)\ dx = 0$, U being the unipotent radical of P.

2) For any cuspidal subgroup $P \neq G$, we have $\int_{U/U\cap\Gamma} f(xu)du = 0$ for almost all $x \in G$.

3) For any maximal cuspidal subgroup P of G, $\int_{U/U\cap\Gamma} f(xu)du = 0$ for almost all $x \in G$.

<u>Proof</u>. The equivalence of 1) and 2) is immediate from the formula

$$\int_{G/U\cap\Gamma} \varphi(x)f(x)dx = \int_{G/U} \varphi(x)d\bar{x} \int_{U/U\cap\Gamma} f(xu)du .$$

Suppose that f satisfies 3). Let P be a cuspidal group $\neq G$. Choose a split component A of P and a subset F of $\Sigma^0(P|A)$ whose complement consists of one element. Then P_F is a maximal cuspidal subgroup and

$$\int_{U/U\cap\Gamma} f(xu)du = \int_{U/U_F(U\cap\Gamma)} d\bar{u} \int_{U_F/U_F\cap\Gamma} f(xuv)dv = 0 .$$

So 3) is equivalent with 2).

<u>Definition</u>. ${}^0L^2(G/\Gamma)$ = the subspace of $L^2(G/\Gamma)$ consisting of the elements which satisfy the equivalent conditions 1), 2), 3) of Lemma 3. ${}^0L^2(G/\Gamma)$ is called the <u>space of cusp forms</u>.

Clearly ${}^0L^2(G/\Gamma)$ is a closed subspace of $L^2(G/\Gamma)$ which is invariant under G. For $\alpha \in C_c(G)$ the operator $\lambda(\alpha)$ on $L^2(G/\Gamma)$ is defined by

$$\lambda(\alpha)f = \int_G \alpha(x)\ \lambda(x)f\ dx = \alpha * f .$$

We denote by ${}^0\lambda(\alpha)$ the restriction of $\lambda(\alpha)$ on ${}^0L^2(G/\Gamma)$ and shall prove:

Theorem 2. (Gelfand and Piatetsky-Shapiro). If $\alpha \in C_c^\infty(G)$, the operator ${}^0\lambda(\alpha)$ is compact. (Langlands has proved this for $\alpha \in C_c^1(G)$).

§3. Inequalities. Statement of the main lemma.

Let (P,A) be a mincuspidal pair, $\mathfrak{S}$ a Siegel domain with respect to (P,A) and Ξ a finite subset of $G_{\mathbb{Q}}$. The following two lemmas have been proved in Borel's Paris lectures.

Lemma 4. There exist constants c_1, c_2 with $c_1 \geq c_2 > 0$ such that

$$c_1\|x\| \geq \inf_{\gamma\in\Gamma}\|x\gamma\| \geq c_2\|x\| \quad \text{for } x \in \mathfrak{S}\Xi .$$

Lemma 5. a) There exist $c \geq 0$, $\lambda \in \mathfrak{a}^*$ such that $\|x\| \leq c\, e^{\lambda(H(x))}$ for $x \in \mathfrak{S}$.
b) Given any $\Lambda \in \mathfrak{a}^*$, there exist c' and N such that $e^{\Lambda(H(x))} \leq c'\|x\|^N$ for $x \in \mathfrak{S}$. (Here H stands for $H_{P|A}$).

Assume now that $G = \mathfrak{S}\Xi\Gamma$. Then the next lemma is an immediate consequence of Lemmas 4 and 5.

Lemma 6. For a function f on G/Γ the following conditions are equivalent.

1) There exist c and r such that $|f(x)| \leq c\|x\|^r$ for $x \in G$.
2) There exist c' and $\lambda \in \mathfrak{a}^*$ such that $|f(x\xi)| \leq c'\, e^{\lambda(H(x))}$ for $x \in \mathfrak{S}$, $\xi \in \Xi$.

Lemma 7. For any $t > 0$ let G_t be the set of $x \in G$ with $\|x\| \leq t$. Then there are numbers c and N such that $\mathrm{vol}(G_t) \leq ct^N$ for $t > 0$.

[See Borel (Paris) or Harish-Chandra (IHES, No. 27)].

From this one deduces immediately:

Lemma 8. If $\Gamma_t = \Gamma \cap G_t$, the number of elements of Γ_t satisfies an inequality

$$[\Gamma_t] \leq c'\, t^N \qquad (t > 0).$$

Lemma 9. Let α be a bounded function on G with compact support. We can choose c and N such that

$$\sum_{\gamma\in\Gamma} |\alpha(x\gamma y)| \leq c\|x\|^N \quad \text{for } x,y \in G .$$

Proof. It is sufficient to prove it for the characteristic function α of a compact set Ω. Then $\sum_{\gamma\in\Gamma} \alpha(x\gamma y) = [\Gamma \cap x^{-1}\Omega y^{-1}]$. If γ and γ_o belong to $\Gamma \cap x^{-1}\Omega y^{-1}$ then $\gamma\gamma_o^{-1} \in \Gamma \cap x^{-1}\Omega\Omega^{-1}x$, so $[\Gamma \cap x^{-1}\Omega y^{-1}] \leq [\Gamma \cap x^{-1}\Omega\Omega^{-1}x]$. Now $\Omega\Omega^{-1}$ is contained in some set G_a (in the notation of Lemma 7) and $x^{-1}\Omega\Omega^{-1}x$ is contained in $G_{b\|x\|^M}$ for some constants b,M. Hence $\sum_{\gamma\in\Gamma} \alpha(x\gamma y) \leq [\Gamma_{b\|x\|^M}] \leq c\,\|x\|^N$ (Lemma 8).

Remark. This proof shows that the same N can be chosen for all α.

Corollary. Let α be a bounded measurable function with compact support on G. We can choose c and N such that

$$|(\alpha * \varphi)(x)| \leq c\,\|x\|^N\|\varphi\|_1 \quad \text{for all } \varphi \in L^1(G/\Gamma),\ x \in G .$$

Proof. $|(\alpha * \varphi)(x)| \leq \int_G |\alpha(xy^{-1})\varphi(y)|\,dy = \int_{G/\Gamma} \sum_\Gamma |\alpha(x\gamma y^{-1})|\,|\varphi(y)|\,dy \leq c\,\|x\|^N\|\varphi\|_1$ where c and N are as in Lemma 9.

The following notations will be used in the sequal.

If $g \in \mathfrak{g}$ = algebra of all left-invariant differential operators on G and if $f \in C^\infty(G)$, then $(gf)(x) = f(x;g)$. For $X \in \mathfrak{g}$, we have $(Xf)(x) = f(x;X) = \left(\frac{d}{dt} f(x \exp tX)\right)_{t=0}$, and put $(\varrho(X)f)(x) = f(X;x) = \left(\frac{d}{dt} f(\exp tX.x)\right)_{t=0}$. Then $[\varrho(X),\varrho(Y)] = -\varrho([X,Y])$ and ϱ can be extended to an antiisomorphism of $\mathfrak{g}$ on the algebra of right-invariant differential operators on G. Notation: $(\varrho(g)f)(x) = f(g;x)$, and sometimes: $\varrho(g)f = g'f$. Finally: $((\varrho(g_1)\circ g_2)f)(x) = ((g_2\circ\varrho(g_1))f)(x) = f(g_1;x;g_2)$.

Fix a mincuspidal pair (P,A) and let $P = MAU$ be the Langlands decomposition. A subset Σ' of Σ is called an <u>ideal</u> in Σ if $\alpha \in \Sigma'$, $\beta \in \Sigma$ and $\alpha + \beta \in \Sigma$ imply $\alpha + \beta \in \Sigma'$. If Σ' is an ideal in Σ, $\mathfrak{u}(\Sigma') = \sum_{\alpha\in\Sigma'} \mathfrak{u}_\alpha$ is an ideal in $\mathfrak{u}$ and the corresponding analytic subgroup $U(\Sigma')$ of G is a normal subgroup of U. We assume that we are given a discrete subgroup Γ_∞ of U and a Siegel domain $\mathfrak{S}$ with respect to (P,A) such that the following conditions are satisfied.

1) $\mathfrak{S}\Gamma_\infty = \mathfrak{S}U$;

2) if Σ' is an ideal in Σ, then $U(\Sigma')/U(\Sigma') \cap \Gamma_\infty$ is compact.

In particular, $U/U \cap \Gamma_\infty$ is compact.

For each $g \in \mathfrak{G}$, $\lambda \in \mathfrak{a}^*$ we define a seminorm $\nu_{g,\lambda}$ on $C^\infty(G)$ by

$$\nu_{g,\lambda}(f) = \sup_{x \in \mathfrak{F}} |f(g;x)|\, e^{\lambda(H(x))}.$$

For any subset F of Σ^0 and $\lambda_0 \in \mathfrak{a}^*$, let $\mathcal{S}_F(\lambda_0)$ be the set of all seminorms $\nu_{g,\lambda}$ with $g \in \mathfrak{G}$, $\lambda \in \mathfrak{a}^*$, $\lambda = \lambda_0$ on $\mathfrak{a}_F$. We define $\mathcal{L}_F(\lambda_0)$ as the space of all functions $f \in C^\infty(G/\Gamma_\infty)$ such that $\nu(f) < \infty$ for $\nu \in \mathcal{S}_F(\lambda_0)$. We topologize $\mathcal{L}_F(\lambda_0)$ by means of the set of seminorms $\mathcal{S}_F(\lambda_0)$ (this topology is not necessarily Hausdorff). Put $\mathcal{L}(\lambda_0) = \mathcal{L}_\emptyset(\lambda_0)$ and $\mathcal{L} = \mathcal{L}_{\Sigma^0}(\lambda_0)$. Then $\mathcal{L}(\lambda_0)$ is defined by the seminorms ν_{g,λ_0} $(g \in \mathfrak{G})$ and $\mathcal{L}$ by the seminorms $\nu_{g,\lambda}$ $(g \in \mathfrak{G}, \lambda \in \mathfrak{a}^*)$. If $F \supset F'$, then $\mathcal{L}_F(\lambda_0) \subset \mathcal{L}_{F'}(\lambda_0)$ and the inclusion map is continuous. $\mathcal{L}_{F_1 \cup F_2}(\lambda_0) = \mathcal{L}_{F_1}(\lambda_0) \cap \mathcal{L}_{F_2}(\lambda_0)$ and the topology of $\mathcal{L}_{F_1 \cup F_2}(\lambda_0)$ is the weakest topology which will make both inclusions $\mathcal{L}_{F_1 \cup F_2}(\lambda_0) \to \mathcal{L}_{F_i}(\lambda_0)$ $(i = 1,2)$ continuous; this follows easily from the trivial remark that $\nu_{g,\mu}(f)^2 \leq \nu_{g,\lambda_1}(f)\nu_{g,\lambda_2}(f)$ if $\mu = \frac{1}{2}(\lambda_1+\lambda_2)$. The topology of $\mathcal{L}_F(\lambda_0)$ is the weakest topology which will make the inclusions $\mathcal{L}_F(\lambda_0) \to \mathcal{L}_\alpha(\lambda_0)$ $(\alpha \in F)$ continuous; here $\mathcal{L}_\alpha(\lambda_0) = \mathcal{L}_{\{\alpha\}}(\lambda_0)$.

For any normal analytic subgroup V of U such that $V/V \cap \Gamma_\infty$ is compact, we define $(\pi_V f)(x) = \int_{V/V\cap\Gamma_\infty} f(xv)\,dv$ $(f \in C(G/\Gamma_\infty),\ x \in G)$ where the Haar measure dv on V is normalized by the condition $\int_{V/V\cap\Gamma_\infty} dv = 1$. Obviously $\pi_V f$ is a continuous function on G which is right-invariant under $V\Gamma_\infty$. Suppose $f \in C^\infty(G/\Gamma_\infty)$. Then

$$\nu_{g,\lambda}(\pi_V f) = \sup_{x\in\mathfrak{F}} \left| \int_{V/V\cap\Gamma_\infty} f(g;xv)\, e^{\lambda(H(xv))}dv \right| \leq \sup_{y\in\mathfrak{F}U} \left| f(g;y)\, e^{\lambda(H(y))} \right| = \nu_{g,\lambda}(f),\quad \text{since}$$

$\mathfrak{F}U = \mathfrak{F}\Gamma_\infty$. So for every F and every λ, π_V defines a continuous mapping of $\mathcal{L}_F(\lambda)$ into itself.

Now we define $\pi_F = \pi_V$ where $V = U_{c_F}$, and $\pi_\alpha = \pi_{\{\alpha\}}$. Finally, put $\pi'_\alpha = 1 - \pi_\alpha$. Now we can state the main lemma of this chapter.

<u>Lemma 10</u>. Let $\alpha \in \Sigma^0$, $\lambda \in \mathfrak{a}^*$. For any $f \in \mathcal{L}(\lambda)$, $\pi'_\alpha f$ belongs to $\mathcal{L}_\alpha(\lambda)$ and the mapping of $\mathcal{L}(\lambda)$ into $\mathcal{L}_\alpha(\lambda)$ defined by π'_α is continuous.

This lemma has as an immediate consequence the following important result.

Corollary. Let ${}^{\circ}\mathcal{L}(\lambda)$ denote the set of all $f \in \mathcal{L}(\lambda)$ such that $\pi_\alpha f = 0$ for all $\alpha \in \Sigma^{\circ}$. Then ${}^{\circ}\mathcal{L}(\lambda) \subset \mathcal{L}$ and the corresponding injection is continuous.

We shall give the proof of the main lemma in §7 and first derive Theorems 1 and 2 from it.

We conclude this paragraph with a formula which will be used later.

Put $f_F = \pi_{c_F} f$, in other words $f_F(x) = \int_{U_F/U_F \cap \Gamma_\infty} f(xu)\,du$, and put ${}^{\circ}f = \left(\prod_{\alpha\in\Sigma^{\circ}} \pi'_\alpha\right) f$ (remark that the π_α commute).

Lemma 11. Let $f \in C(G/\Gamma_\infty)$. Then $f = {}^{\circ}f - \sum_{F\neq\Sigma^{\circ}} (-1)^{\operatorname{rank} F} f_F$, where rank F = = rank $P_F = [\Sigma^{\circ}] - [F]$.

Proof. It is easily seen that $\pi_F = \prod_{\alpha\in F} \pi_\alpha$. A trivial computation proves the formula.

§4. The space of cusp forms.

Fix a mincuspidal pair (P,A), and choose a Siegel domain $\mathcal{J} = KA_t\omega$ with respect to (P,A) and a finite subset Ξ of $G_{\mathbb{Q}}$ such that $G = \mathcal{J}\,\Xi\,\Gamma$. Put $\Gamma_\infty = U \cap \bigcap_{\xi\in\Xi} {}^{\xi}\Gamma$. By taking ω large enough we may suppose that $\omega\Gamma_\infty = \omega U$, because U/Γ_∞ is compact. Then the two conditions of §3 on $\mathcal{J}$ and Γ_∞ are fulfilled. We use now the notation of §3. First we introduce some function spaces. In the following definitions we set $f_\xi(x) = f(x\xi)$.

$\mathcal{A}^\infty(G/\Gamma,\lambda) = \{f \in C^\infty(G/\Gamma) : f_\xi \in \mathcal{L}(\lambda) \text{ for } \xi \in \Xi\}$.

$\mathcal{A}(G/\Gamma,\lambda) = \{f \in C(G/\Gamma) : \sup_{x\in\mathcal{J},\xi\in\Xi} |f_\xi(x)|\, e^{\lambda(H(x))} < \infty\}$.

$\mathcal{A}(G/\Gamma) = \{f \in C(G/\Gamma) : \exists r \text{ such that } \sup_{x\in G} |f(x)|\, \|x\|^{-r} < \infty\}$.

$\mathcal{A}^\infty(G/\Gamma) = \{f \in C^\infty(G/\Gamma) : \exists r \text{ such that for any } g \in \mathcal{G} \text{ one has } \sup_{x\in G} |f(g;x)|\, \|x\|^{-r} < \infty\}$.

For $f \in C^\infty(G/\Gamma)$, put $\nu'_{g,\lambda}(f) = \sup_{\xi\in\Xi} \nu_{g,\lambda}(f_\xi)$. Then $f \in \mathcal{A}^\infty(G/\Gamma,\lambda)$ means $\nu'_{g,\lambda}(f) < \infty$ for all $g \in \mathcal{G}$. It is clear that $\mathcal{A}(G/\Gamma,\lambda) \subset \mathcal{A}(G/\Gamma)$ and $\mathcal{A}^\infty(G/\Gamma,\lambda) \subset \mathcal{A}^\infty(G/\Gamma)$ (cf. Lemma 6). We denote by ${}^{\circ}\mathcal{A}^\infty(G/\Gamma,\lambda)$ the space of all $f \in \mathcal{A}^\infty(G/\Gamma,\lambda)$ such that $f_{P'}(x) = \int_{U'/U'\cap\Gamma} f(xu)\,du = 0$ for every cuspidal subgroup P'

of G with unipotent radical $U' \neq 1$; analogous definition for ${}^{\circ}\mathcal{A}(G/\Gamma,\lambda)$, ${}^{\circ}\mathcal{A}(G/\Gamma)$, ${}^{\circ}\mathcal{A}^{\infty}(G/\Gamma)$.

Let $f \in C(G/\Gamma)$. Then, for $\alpha \in \Sigma^{o}$, $\xi \in \Xi$, we have

$$\pi_{\alpha} f_{\xi}(x) = \int_{U_F/U_F \cap \Gamma_{\infty}} f(xu\xi)\,du = \int_{U'/U' \cap \Gamma} f(x\xi u')\,du' ,$$

where $F = {}^{c}\{\alpha\}$, $U' = {}^{\xi^{-1}}U_F$. This shows that it is equivalent to say that $\pi_{\alpha} f_{\xi} = 0$ for all $\alpha \in \Sigma^{o}$, $\xi \in \Xi$ or to say that $f_{P'} = 0$ for all P' of the form ${}^{\xi^{-1}}P_F$ with $\xi \in \Xi$, $F \subset \Sigma^{o}$, rank $F = 1$. But, since $G_{\mathbb{Q}} = P_{\mathbb{Q}} \Xi \Gamma$, every maximal cuspidal subgroup of G is of the form ${}^{\gamma\xi^{-1}}P_F$ with $\gamma \in \Gamma$, $\xi \in \Xi$, $F \subset \Sigma^{o}$, rank $F = 1$. Moreover $f_{\gamma P'}(x) = f_{P'}(x\gamma)$. Thus we see that, in order that $f_{P'} = 0$ for every maximal cuspidal subgroup P' of G, it is necessary and sufficient that $\pi_{\alpha} f_{\xi} = 0$ for all $\alpha \in \Sigma^{o}$, $\xi \in \Xi$. This assertion remains true when one replaces the word "maximal" by "proper" (cf. Lemma 3). In particular, if $f \in \mathcal{A}^{\infty}(G/\Gamma,\lambda)$, then $f \in {}^{\circ}\mathcal{A}^{\infty}(G/\Gamma,\lambda)$ if and only if $f_{\xi} \in {}^{\circ}\mathcal{L}(\lambda)$ for all $\xi \in \Xi$. The corollary of Lemma 10 tells us now that, if $f \in {}^{\circ}\mathcal{A}^{\infty}(G/\Gamma,\lambda)$, then $f_{\xi} \in \mathcal{L}$ for all $\xi \in \Xi$ (d'une manière plus imagée: if a slowly increasing function in $C^{\infty}(G/\Gamma)$ is a cusp function, then it is rapidly decreasing). Moreover, we have:

<u>Lemma 12</u>. Fix $\lambda_o \in \mathfrak{a}^*$. For any given $g \in \mathfrak{G}$, $\lambda \in \mathfrak{a}^*$, there exists a continuous seminorm ν on $\mathcal{L}(\lambda_o)$ such that

$$\nu'_{g,\lambda}(f) \leq \nu'(f) = \sup_{\xi \in \Xi} \nu(f_{\xi}) \quad \text{for all } f \in {}^{\circ}\mathcal{A}^{\infty}(G/\Gamma,\lambda_o) .$$

<u>Proof</u>. One has merely to apply the corollary of Lemma 10 to ${}^{\circ}\mathcal{L}(\lambda_o)$.

<u>Lemma 13</u>. Let $\alpha \in C_c^{\infty}(G)$ and $f \in {}^{o}L^2(G/\Gamma)$. Then $\alpha * f \in {}^{\circ}\mathcal{A}^{\infty}(G/\Gamma)$. Moreover, given $g \in \mathfrak{G}$, $\lambda \in \mathfrak{a}^*$, we can choose a continuous seminorm σ on $C_c^{\infty}(G)$ such that

$$\nu'_{g,\lambda}(\alpha * f) \leq \sigma(\alpha)\, \|f\|_2 \quad \text{for all } \alpha \in C_c^{\infty}(G),\ f \in {}^{o}L^2(G/\Gamma).$$

<u>Proof</u>. Fix a compact set $\Omega \subset G$. There are constants c and N such that

$$|(\alpha * f)(x)| \leq c\|\alpha\|_{\infty} \|x\|^{N} \|f\|_1 \quad \text{for all } \alpha \in C_{\Omega}^{\infty}(G),\ f \in L^1(G/\Gamma)$$

(apply the corollary of Lemma 9 to the characteristic function of Ω). Since G/Γ has finite volume, we have $\|f\|_1 \leq c_1\|f\|_2$ for $f \in L^2(G/\Gamma)$. If $g \in \mathfrak{g}$, then $(\alpha * f)(g;x) = (g'\alpha * f)(x)$, and we get from the above inequalities $|(\alpha * f)(g;x)| \leq cc_1\|g'\alpha\|_\infty\|x\|^N\|f\|_2$ $(\alpha \in C^\infty_\Omega(G),\ f \in L^2(G/\Gamma),\ g \in \mathfrak{g})$. This implies already that $\alpha * f \in \mathcal{A}^\infty(G/\Gamma)$. Moreover, $(\alpha * f)_{P'} = \alpha * f_{P'}$, so that $\alpha * f \in {}^0\mathcal{A}^\infty(G/\Gamma)$ if $f \in {}^0L^2(G/\Gamma)$. By Lemma 5 we can choose $\lambda_0 \in \mathfrak{a}^*$ such that $\sup_{x\in\mathfrak{S}} \|x\|^N e^{\lambda_0(H(x))} < \infty$. Then $\nu'_{g,\lambda_0}(\alpha * f) < \infty$ for all g, hence $\alpha * f \in {}^0\mathcal{A}^\infty(G/\Gamma,\lambda_0)$. By Lemma 12, for given $g \in \mathfrak{g}$, $\lambda \in \mathfrak{a}^*$, there exists a continuous seminorm ν on $\mathcal{A}^\infty(G/\Gamma,\lambda_0)$ such that $\nu'_{g,\lambda}(\alpha * f) \leq \nu(\alpha * f)$. Now, for $\alpha \in C^\infty_\Omega(G)$, $f \in {}^0L^2(G/\Gamma)$ we have

$$\nu'_{g,\lambda}(\alpha * f) \leq \nu(\alpha * f) \leq \sum_i \nu'_{g_i,\lambda_0}(\alpha * f) \leq cc_1\sum\|g_i'\alpha\|_\infty \sup_{\substack{x\in\mathfrak{S}\\ \xi\in\Xi}}\|x\xi\|^N e^{\lambda_0(H(x))}\|f\|_2 \leq$$

$$\leq c_2 \sum\|g_i'\alpha\|_\infty\|f\|_2$$

with certain elements g_i of $\mathfrak{g}$ (finite in number). The second assertion of the lemma follows from this.

Proof of Theorem 2.

Let $\alpha \in C^\infty_c(G)$. As a consequence of Lemma 13 we have

$$\sup_{x\in G} |(\alpha * f)(x)| \leq c\|f\|_2 \qquad (f \in {}^0L^2(G/\Gamma))$$

where c is independent of f. Hence $(\alpha * f)(x) = (k_x,f)$ with $k_x \in {}^0L^2(G/\Gamma)$ and $\|k_x\|_2 \leq c$. If $k(x,y) = k_x(y)$, then

$$\iint|k(x,y)|^2\,dx\,dy \leq \int \|k_x\|_2^2\,dx < \infty \quad ,$$

so ${}^0\lambda(\alpha)$ is an operator of Hilbert-Schmidt type and therefore compact.

Question. Is ${}^0\lambda(\alpha)$ of trace class ?

Let σ, resp. χ, be a representation of K, resp. $\mathfrak{z}$, on a finite-dimensional space V such that $\sigma(k)$ commutes with $\chi(z)$ $(k \in K,\ z \in \mathfrak{z})$. The space $\mathcal{A}(G/\Gamma,\sigma,\chi)$ has been defined in §2 ; let ${}^0\mathcal{A}(G/\Gamma,\sigma,\chi)$ be the space of all $f \in \mathcal{A}(G/\Gamma,\sigma,\chi)$ such that $f_{P'}(x) = \int_{U'/U'\cap\Gamma} f(xu)\,du = 0$ for every cuspidal group $P' \neq G$. We are going to prove that the functions of ${}^0\mathcal{A}(G/\Gamma,\sigma,\chi)$ are square-integrable.

Lemma 14. $\mathcal{A}(G/\Gamma,\sigma,\chi) \subset \mathcal{A}^{\infty}(G/\Gamma) \otimes V$.

Proof. Let $f \in \mathcal{A}(G/\Gamma,\sigma,\chi)$. Then f is an analytic function which is K-finite and $\mathfrak{Z}$-finite, hence there exists a function $\alpha \in C_c^{\infty}(G)$ such that $f = \alpha * f$ (see Harish-Chandra, Acta Math. 116, p. 18). For $g \in \mathfrak{G}$ we have $f(g;x) = (\alpha * f)(g;x) =$ $= (g'\alpha * f)(x) = \int_G g'\alpha(y)\, f(y^{-1}x)\, dy$.

Choose c and r such that $|f(x)| \leq c\|x\|^r$ for all $x \in G$. Then

$$|f(g;x)| \leq c\|x\|^r \int_G |g'\alpha(y)|\,\|y^{-1}\|^r\, dy = c(g)\ \|x\|^r \quad (g \in \mathfrak{G},\ x \in G).$$

Lemma 15. If $\varphi \in \mathcal{A}(G/\Gamma)$ and $f \in {}^{\circ}\mathcal{A}^{\infty}(G/\Gamma)$, then $\int_{G/\Gamma} |\varphi(x)\, f(x)|\, dx < \infty$.

Proof. If $\mathcal{Y}$ and Ξ are as in the beginning of this paragraph, we have $G = \mathcal{Y}\Xi\Gamma$ and it is sufficient to prove that $\int_{\mathcal{Y}} |\varphi(x\xi)\, f(x\xi)|\, dx < \infty$ for every $\xi \in \Xi$. By lemma 6, φ is in some space $\mathcal{A}(G/\Gamma,\lambda_1)$; also f is in some space ${}^{\circ}\mathcal{A}^{\infty}(G/\Gamma,\lambda_0)$. But then f is in any space $\mathcal{A}(G/\Gamma,\lambda)$ (Lemma 12), so φf is in any space $\mathcal{A}(G/\Gamma,\lambda)$, in particular it is in $\mathcal{A}(G/\Gamma,0)$, i.e., $\varphi(x\xi)f(x\xi)$ is bounded for $x \in \mathcal{Y}$, $\xi \in \Xi$. It is well known that $\mathcal{Y}$ has finite volume...

Corollary. ${}^{\circ}\mathcal{A}^{\infty}(G/\Gamma) \subset {}^{\circ}L^2(G/\Gamma)$; we have even more: every function in ${}^{\circ}\mathcal{A}^{\infty}(G/\Gamma)$ is bounded.

Lemma 16. ${}^{\circ}\mathcal{A}(G/\Gamma,\sigma,\chi) \subset {}^{\circ}L^2(G/\Gamma) \otimes V$.

This follows immediately from Lemma 14 and the corollary of Lemma 15.

Let $\mathcal{E}_K$ denote the set of equivalence classes of finite-dimensional irreducible representations of K . For $\vartheta \in \mathcal{E}_K$ we define the space ${}^{\circ}\mathcal{A}(G/\Gamma,\vartheta,\chi)$ as follows. Choose a representation (σ,V) belonging to the equivalence class ϑ . Then the space ${}^{\circ}\mathcal{A}(G/\Gamma,\vartheta,\chi)$ is spanned by the functions $x \to (v,f(x))$ with $f \in {}^{\circ}\mathcal{A}(G/\Gamma,\sigma,\chi)$, $v \in V$ [we have supplied V with a scalar product (v,w) which we may suppose, whenever that is convenient, to be such that σ is a unitary representation].

Let $\mathcal{H}$ be the set of all characters of $\mathfrak{Z}$, i.e., of all homomorphisms $\mathfrak{Z} \to C$ which map 1 into 1.

Theorem 3. The space $\sum_{\vartheta \in \mathcal{E}_K,\chi\in\mathcal{H}} {}^{\circ}\mathcal{A}(G/\Gamma,\vartheta,\chi)$ is dense in ${}^{\circ}L^2(G/\Gamma)$.

Proof. From Theorem 2 it follows that ${}^{o}L^2(G/\Gamma) = \mathrm{Cl}(\Sigma \mathcal{H}_i)$ where the $\mathcal{H}_i$ are closed invariant irreducible subspaces which are mutually orthogonal. Fix an $\mathcal{H}_i$ and call π the representation of G on $\mathcal{H}_i$. For $\vartheta \in \mathcal{E}_K$ define $E_\vartheta = d(\vartheta) \int_K \overline{\xi_\vartheta(k)}\, \pi(k)\, dk$ where $d(\vartheta)$ is the degree of ϑ and ξ_ϑ the character of ϑ. Then $\mathcal{H}_i = \mathrm{Cl}(\sum_{\vartheta \in \mathcal{E}_K} \mathcal{H}_{i,\vartheta})$, with $\mathcal{H}_{i,\vartheta} = E_\vartheta \mathcal{H}_i$, and $\dim \mathcal{H}_{i,\vartheta} < \infty$. Let χ be the infinitesimal character of π, i.e., we have $z\varphi = \varphi\chi(z)$ $(\varphi \in \mathcal{H}_i,\ z \in \mathfrak{Z})$. If $\varphi \in \mathcal{H}_{i,\vartheta}$, then φ is K-finite and is eigenfunction of $\mathfrak{Z}$, so $\varphi = \alpha * \varphi$ for some $\alpha \in C_c^\infty(G)$. By Lemma 13 we have then $\varphi \in {}^{o}\mathcal{A}^\infty(G/\Gamma)$, hence $\varphi \in {}^{o}\mathcal{A}(G/\Gamma,\vartheta,\chi)$, so that $\mathcal{H}_{i,\vartheta} \subset {}^{o}\mathcal{A}(G/\Gamma,\vartheta,\chi)$ and the theorem is proved.

Remarks. 1. For each i the number of j such that $\mathcal{H}_j$ is equivalent to $\mathcal{H}_i$ is finite (consequence of Theorem 2).

2. Using representation theory we know that for fixed $\chi \in \mathcal{X}$ and $\vartheta \in \mathcal{E}_K$ there are only finitely many i such that 1) $\mathcal{H}_{i,\vartheta} \neq o$ and 2) $\mathcal{H}_i$ has the infinitesimal character χ. Because $\dim \mathcal{H}_{i,\vartheta} < \infty$, this implies that $\dim {}^{o}\mathcal{A}(G/\Gamma,\vartheta,\chi) < \infty$. However we shall give another proof of that fact (see §6).

§5. A theorem of Langlands.

As a consequence of Theorem 3 we prove here a theorem whose importance will appear in Chapter II.

Let σ be a finite-dimensional representation of K. If $f \in \mathcal{A}(G/\Gamma,\sigma)$ and if $P = MAU$ is a cuspidal subgroup, then $|f_P(x)| \leq \int_{U/U\cap\Gamma} |f(xu)|\,du \leq c\|x\|^r$ for some c,r. For every $a \in A$ the function $m \to f_P(ma)$ on M belongs to $\mathcal{A}(M/\Gamma_M,\sigma_M)$.

Theorem 4. (Langlands). Suppose the function f in $\mathcal{A}(G/\Gamma,\sigma)$ has the following property. For any cuspidal pair (P,A), for any $\varphi \in {}^{o}\mathcal{A}(M/\Gamma_M,\sigma_M,\chi)$, where χ is any character of $\mathfrak{Z}_M$, and for any $a \in A$, we have

$$\int_{M/\Gamma_M} (\varphi(m), f_P(ma))\, dm = o .$$

Then $f = o$.

Remarks. 1. Notice that the cuspidal pair $(G,1)$ is included in the hypotheses.

2. The integral in the assertion of Theorem 4 exists always, as one sees from Lemma 15.

Proof. We use induction on $\mathrm{rank}_{\mathbb{Q}}(G)$. Assume first that $\mathrm{rank}_{\mathbb{Q}}(G) = 0$. Then $\mathcal{A}(G/\Gamma,\sigma) = {}^{0}\mathcal{A}(G/\Gamma,\sigma)$. For any $\alpha \in C_c^{\infty}(G)$ such that $\alpha(kxk^{-1}) = \alpha(x)$ $(k \in K,\ x \in G)$ we have $(\alpha * f)(kx) = \sigma(k)(\alpha * f)(x)$, and as in the proof of Lemma 14 one sees that $\alpha * f \in \mathcal{A}^{\infty}(G/\Gamma) \otimes V$; so $\alpha * f$ is a function of ${}^{0}\mathcal{A}^{\infty}(G/\Gamma,\sigma)$. The latter space is contained in ${}^{0}L^{2}(G/\Gamma,\sigma)$ (Corollary of Lemma 15) and, from Theorem 3, ${}^{0}L^{2}(G/\Gamma,\sigma) = \mathrm{Cl}(\sum_{\chi\in\mathcal{X}} {}^{0}\mathcal{A}(G/\Gamma,\sigma,\chi))$. By hypothesis $\int_{G/\Gamma}(\varphi,f)dx = 0$ for all $\varphi \in \sum_{\chi} {}^{0}\mathcal{A}(G/\Gamma,\sigma,\chi)$, hence $\int_{G/\Gamma}(\alpha * f,f)dx = 0$. Let now α tend to the Dirac distribution with support $\{1\}$, then it follows that $\int_{G/\Gamma}(f,f)dx = 0$, and $f = 0$.

Suppose now that $\mathrm{rank}_{\mathbb{Q}}(G) = l \geq 1$. Let P be a proper cuspidal subgroup of G, $P = MAU$, and let $*P = *M*A*U$ be a cuspidal subgroup of M. If $P' = M'A'U'$ is the corresponding cuspidal subgroup of G contained in P (Lemma 2), then the following facts are easily verified. $\pi_{P'|M'} = \pi_{*P|*M} \circ \pi_{P|M}$ on ${}^{0}P'$, and from this $\Gamma_{M'} = (\Gamma_{M})_{*M}$. If $g(m) = f_P(ma)$ $(m \in M,\ a \in A)$, then

$$g_{*P}(m) = \int_{*U/*U \cap \Gamma_M} g(m*u)d*u = f_{P'}(ma) .$$

For any $\varphi \in {}^{0}\mathcal{A}(M'/\Gamma_{M'},\sigma_{M'},\chi)$ with $\chi \in \mathcal{X}_{M'}$ and any $*a \in *A$ we have

$$\int_{*M/\Gamma_{*M}} (\varphi(*m),\ g_{*P}(*m*a))\ d*m = \int_{M'/\Gamma_{M'}} (\varphi(m'),f_{P'}(m'*aa))dm' = 0 .$$

Since $g \in \mathcal{A}(M/\Gamma_M,\sigma_M)$ and $\mathrm{rank}_{\mathbb{Q}}(M) = l\text{-rank } P < l$, we have by induction $g = 0$. So $f_P = 0$ for all $P \neq G$, that is $f \in {}^{0}\mathcal{A}(G/\Gamma,\sigma)$. By the same argument as in the case $\mathrm{rank}_{\mathbb{Q}}(G) = 0$ one proves that $f = 0$.

§6. Proof of Theorem 1.

Lemma 17. (Godement). Let X be a locally compact space and μ a positive measure on X such that $\mu(X) = 1$. Let $\mathcal{H}$ be a closed subspace of $L^{2}(X,\mu)$. Suppose that every function f in $\mathcal{H}$ is essentially bounded on X. Then $\dim \mathcal{H} < \infty$.

Proof. (Hörmander). Since $\|f\|_2 \leq \|f\|_\infty$ for $f \in \mathcal{H}$, the identical mapping of $\mathcal{H}$ (supplied with the norm $\| \ \|_\infty$) on $\mathcal{H}$ (supplied with the norm $\| \ \|_2$) is continuous. By the closed-graph theorem this mapping is an homeomorphism, so $\|f\|_\infty \leq c \|f\|_2$ for $f \in \mathcal{H}$. Let $f_1, \ldots, f_n$ be orthonormal elements of $\mathcal{H}$. Then

$$(*) \qquad \left| \sum_{1 \leq j \leq n} a_j f_j(x) \right| \leq c\Big(\sum_{1 \leq j \leq n} |a_j|^2 \Big)^{1/2}$$

for almost all x if $a = (a_1, \ldots, a_n) \in \mathbf{C}^n$ is fixed. Hence this is true for almost all x and all a in some countable dense subset D of $\mathbf{C}^n$. But, if $(*)$ holds for a certain x and all $a \in D$, then it holds for x and all $a \in \mathbf{C}^n$. So the inequality $(*)$ holds almost everywhere for arbitrary $a \in \mathbf{C}^n$. Taking $a_j = \overline{f_j(x)}$ we get

$$\sum_{1 \leq j \leq n} |f_j(x)|^2 \leq c^2 \qquad \text{(almost everywhere)},$$

and integration gives $n \leq c^2$. Thus $\dim \mathcal{H} \leq c^2$.

Lemma 18. $\dim {}^0\mathcal{A}(G/\Gamma, \sigma, \chi) < \infty$.

Proof. Take in Lemma 17, $X = G/\Gamma$, $\mu = dx$, $\mathcal{H} = {}^0\mathcal{A}(G/\Gamma, \sigma, \chi)$ (that the functions of this space have values in a finite-dimensional vector space instead of $\mathbf{C}$ does not make any difference). The only condition on $\mathcal{H}$ that we have still to verify is that it is closed in $L^2(G/\Gamma) \otimes V$. If $\varphi_n \in \mathcal{H}$ and $\varphi_n \to \varphi$ in $L^2(G/\Gamma) \otimes V$, then, of course, $\varphi \in {}^0L^2(G/\Gamma, \sigma)$, Moreover, $\varphi_n \to \varphi$ in L^2 implies $\varphi_n \to \varphi$ in L^1 and this implies $\varphi_n \to \varphi$ as distributions; therefore, derivation being a continuous operation on distributions, $z\varphi_n \to z\varphi$ for $z \in \mathfrak{Z}$, and from this it follows that $z\varphi = \varphi\chi(z)$. So $\varphi \in {}^0L^2(G/\Gamma, \sigma, \chi)$. Now $\varphi = \alpha * \varphi$ for some function $\alpha \in C_c^\infty(G)$, so $\varphi \in {}^0\mathcal{A}(G/\Gamma, \sigma, \chi)$ (Lemma 13) and $\mathcal{H}$ is closed.

Remark. ${}^0\mathcal{A}(G/\Gamma, \sigma, \chi) = {}^0L^2(G/\Gamma, \sigma, \chi)$.

Before we give the proof of Theorem 1 we recall some facts about the universal enveloping algebra.

Let $P = MAU$ be a cuspidal subgroup of G. Put $M_1 = MA$, $\mathfrak{m}_1 = \mathfrak{m} + \mathfrak{a}$. Let $\mathfrak{M}_1, \mathfrak{M}, \mathfrak{A}$ be the enveloping algebras of $\mathfrak{m}_{1,c}, \mathfrak{m}_c, \mathfrak{a}_c$; they are subalgebras of

$\mathfrak{g}$ and we have $\mathfrak{m}_1 = \mathfrak{m}\mathfrak{a}$. Let $\mathfrak{z}_1$ resp. $\mathfrak{z}_M$ denote the center of $\mathfrak{m}_1$ resp. $\mathfrak{m}$; then $\mathfrak{z}_1 = \mathfrak{z}_M \mathfrak{a}$. Now there is a canonical homomorphism μ of $\mathfrak{z}$ into $\mathfrak{z}_1$. This homomorphism is defined as follows. Consider a Cartan subalgebra $\mathfrak{j}$ of $\mathfrak{m}_1$. Then $\mathfrak{j}$ is also a Cartan subalgebra of $\mathfrak{g}$. Let W, resp. W_1, be the Weyl group of $(\mathfrak{g},\mathfrak{j})$, resp. $(\mathfrak{m}_1,\mathfrak{j})$; W_1 is a subgroup of W, and W acts on the symmetrical algebra $S(\mathfrak{j}_c)$ of $\mathfrak{j}_c$. Let $I(\mathfrak{j}_c)$, resp. $I_1(\mathfrak{j}_c)$, be the algebra of invariants of W, resp. W_1, in $S(\mathfrak{j}_c)$. We have canonical isomorphisms $\gamma : \mathfrak{z} \to I(\mathfrak{j}_c)$ and $\gamma_1 : \mathfrak{z}_1 \to I_1(\mathfrak{j}_c)$. Now $I(\mathfrak{j}_c)$ is a subalgebra of $I_1(\mathfrak{j}_c)$ and $I_1(\mathfrak{j}_c)$ is a free module over $I(\mathfrak{j}_c)$ of finite rank (the rank is equal to $(W:W_1)$). By means of the isomorphisms γ and γ_1 one defines an injective homomorphism $\mu : \mathfrak{z} \to \mathfrak{z}_1$; this homomorphism does not depend on $\mathfrak{j}$ and makes $\mathfrak{z}_1$ a free $\mathfrak{z}$-module of finite rank.

Define $\varrho(X) = \frac{1}{2}\,\mathrm{tr}_{\mathfrak{n}}(\mathrm{ad}\,X)$, $X' = X + \varrho(X)$ for $X \in \mathfrak{m}_1$; here $X' \in \mathfrak{M}_1$. The mapping $X \longmapsto X'$ extends to an automorphism of $\mathfrak{M}_1$. We have the following result: For any $z \in \mathfrak{Z}$, we have $z - \mu(z)' \in \mathfrak{G}\mathfrak{n}$ and

$$\mu(z)' = d^{-1}\mu(z)\circ d$$

as differential operators on M_1; here d is the function on M_1 defined by

$$d(m) = \left|\det_{\mathfrak{n}}(\mathrm{Ad}(m))\right|^{1/2} \qquad (m \in M_1).$$

[For the proof, see: Harish-Chandra, Annals of Math., 83, p. 95-96].
Observe that $d(ma) = e^{\varrho(\log a)}$ if $m \in M$, $a \in A$.

Proof of Theorem 1.

We use induction on $\mathrm{rank}_{\mathbb{Q}}(G)$. The theorem has already been proved in the case $\mathrm{rank}_{\mathbb{Q}}(G) = 0$; so, suppose that this rank is strictly positive. Let P be any proper cuspidal subgroup of G and $P = MAU$ a Langlands decomposition. Define, for $f \in \mathcal{A}(G/\Gamma,\sigma,\chi)$, the function $\pi_P f$ on $M_1 = MA$ by $(\pi_P f)(m) = d(m)\, f_P(m)$ $(m \in M_1)$, where f_P has the usual meaning and d is as above. Let $z \in \mathfrak{Z}$; then, in the above notation, $z = \mu(z)' + \Sigma\, g_i X_i$ with certain $g_i \in \mathfrak{G}$, $X_i \in \mathfrak{n}$. So $zf = \varrho(z)f =$ $= \varrho(\mu(z)')f + \Sigma\varrho(g_i X_i)f$ (where this time ϱ denotes the antiisomorphism of $\mathfrak{G}$ on the algebra of right-invariant differential operators), and from this one finds without trouble that $\pi_P(zf) = \mu(z)\pi_P f$.

Let $\mathfrak{U}$ be the kernel of χ and $\mathfrak{U}_1$ the ideal of $\mathfrak{Z}_1$ generated by $\mu(\mathfrak{U})$. Obviously $\mathfrak{U}_1$ is of finite codimension in $\mathfrak{Z}_1$. Put $\mathfrak{Z}_1^* = \mathfrak{Z}_1/\mathfrak{U}_1$, denote by ζ^* the canonical image in $\mathfrak{Z}_1^*$ of the element ζ of $\mathfrak{Z}_1$, and by $\mathfrak{Z}_1^{**}$ the vector space dual to $\mathfrak{Z}_1^*$. Choose $\zeta_1,\dots,\zeta_r$ in $\mathfrak{Z}_1$ such that $\zeta_1^*,\dots,\zeta_r^*$ is a base for $\mathfrak{Z}_1^*$ and let $\zeta_1^{**},\dots,\zeta_r^{**}$ be the dual base for $\mathfrak{Z}_1^{**}$. For any $\zeta \in \mathfrak{Z}_1$ we have by elementary algebra $\sum_{1\le i\le r}(\zeta\zeta_i^*)\otimes\zeta_i^{**} = \sum_{1\le i\le r}\zeta_i^*\otimes(\zeta_i^{**}\zeta)$ in $\mathfrak{Z}_1^*\otimes_C\mathfrak{Z}_1^{**}$. For $f\in\mathcal{A}(G/\Gamma,\sigma,\chi)$, put $\varphi_f = \sum_{1\le i\le r}(\zeta_i\pi_P f)\otimes\zeta_i^{**}$; φ_f is a function on M_1 with values in $V\otimes_C\mathfrak{Z}_1^{**} = V_1$. If $u\in\mathfrak{U}$, then $\mu(u)\pi_P f = \pi_P(uf) = o$, so $u_1\pi_P f = o$ for $u_1\in\mathfrak{U}_1$, which enables us to define $\zeta^*\pi_P f$ for $\zeta^*\in\mathfrak{Z}_1^*$. Now there is obviously a linear function on $\mathfrak{Z}_1^*\otimes\mathfrak{Z}_1^{**}$ whose values are functions $M_1\to V_1$ and which assigns to an element $\zeta^*\otimes\zeta^{**}$ the function $(\zeta^*\pi_P f)\otimes\zeta^{**}$. From the elementary fact mentioned above it follows that $\Sigma(\zeta\zeta_i\pi_P f)\otimes\zeta_i^{**} = \Sigma(\zeta_i\pi_P f)\otimes(\zeta_i^{**}\zeta)$, i.e., if we define the representation χ_1 of $\mathfrak{Z}_1$ on V_1 by $(v\otimes\zeta^{**})\chi_1(\eta) = v\otimes\zeta^{**}\eta$ ($v\in V$, $\zeta^{**}\in\mathfrak{Z}_1^{**}$, $\eta\in\mathfrak{Z}_1$), then $\zeta\varphi_f = \varphi_f\chi_1(\zeta)$ for $\zeta\in\mathfrak{Z}_1$.

In particular, $H\varphi_f = \varphi_f\chi_1(H)$ for $H\in\mathfrak{A}$. Hence $\varphi_f(ma) = \varphi_f(m)e^{\chi_1(\log a)}$ for $m\in M$, $a\in A$. Denote by ψ_f the restriction of φ_f on M and by χ_M the restriction of χ_1 on $\mathfrak{Z}_M$. Keeping in mind that $\mathcal{A}(G/\Gamma,\sigma,\chi)\subset\mathcal{A}^\infty(G/\Gamma)\otimes V$ one sees that $\psi_f\in\mathcal{A}(M/\Gamma_M,\sigma_M\otimes 1,\chi_M)$. Moreover, it is clear that $\psi_f = o$ is equivalent to $\pi_P f = o$. So, if $\mathcal{A}_P$ is the kernel of π_P in $\mathcal{A} = \mathcal{A}(G/\Gamma,\sigma,\chi)$, then $f\longmapsto\psi_f$ induces a linear injection $\mathcal{A}/\mathcal{A}_P\to\mathcal{A}(M/\Gamma_M,\sigma_M\otimes 1,\chi_M)$. By the induction hypothesis the latter space has finite dimension, hence $\dim\mathcal{A}/\mathcal{A}_P<\infty$.

Let $P_1,\dots,P_s$ be a set of representatives for the conjugacy classes with respect to Γ of all maximal cuspidal subgroups of G .

Since $f_P(x) = f_{P_i}(x\gamma)$ if $P = {}^\gamma P_i$, $\gamma\in\Gamma$, we have $\bigcap_{1\le i\le s}\mathcal{A}_{P_i} = {}^0\mathcal{A}(G/\Gamma,\sigma,\chi) = {}^0\mathcal{A}$. Now $\dim\mathcal{A}/\cap\mathcal{A}_{P_i}\le\Sigma\dim\mathcal{A}/\mathcal{A}_{P_i}<\infty$ after what we have proved above and $\dim{}^0\mathcal{A}<\infty$ after Lemma 18, hence $\dim\mathcal{A}<\infty$.

§7. Proof of the main lemma.

We place ourselves in the situation of the latter part of §3 and shall prove Lemma 10.

Fix $\alpha \in \Sigma^o$ and $\lambda \in \mathfrak{N}^*$. We define a subset Σ_α of Σ as follows. Let $\gamma = \sum_{\beta \in \Sigma^o} m(\beta)\beta$ be a root; then $\gamma \in \Sigma_\alpha$ if and only if $m(\alpha) \neq o$. If $\beta_1 > \beta_2 > \ldots > \beta_s$ are the roots in Σ_α, then $\{\beta_1, \ldots, \beta_i\}$ is an ideal in Σ $(o \leq i \leq s)$. Put $\mathfrak{n}_i = \sum_{1 \leq j \leq i} \mathfrak{n}_{\beta_j}$ and let U_i be the analytic subgroup of G corresponding to $\mathfrak{n}_i$; U_i is a normal subgroup of U and $U_i/U_i \cap \Gamma_\infty$ is compact. We have $U_o = \{1\}$ and $U_s = U_{c\{\alpha\}}$, so $\pi_{U_o} = 1$ and $\pi_{U_s} = \pi_\alpha$.

Lemma 19. For any $f \in \mathcal{L}(\lambda)$, $f - \pi_{U_i} f$ belongs to $\mathcal{L}_\alpha(\lambda)$ and the mapping $f \longmapsto f - \pi_{U_i} f$ of $\mathcal{L}(\lambda)$ into $\mathcal{L}_\alpha(\lambda)$ is continuous $(o \leq i \leq s)$.

Lemma 10 is the case $i = s$ of Lemma 19. We prove Lemma 19 by induction on i. It is true for $i = o$.

Fix $i > o$. Clearly $[\mathfrak{n}, \mathfrak{n}_i] \subset \mathfrak{n}_{i-1}$; in particular, $\mathfrak{n}_i^* = \mathfrak{n}_i/\mathfrak{n}_{i-1}$ is abelian. Put $\Gamma_i = U_i \cap \Gamma_\infty$. If $U_{i-1} = C\Gamma_{i-1}$ with a compact set C, then $U_{i-1}\Gamma_i = C\Gamma_i$, hence $U_{i-1}\Gamma_i$ is closed in U_i. So $U_i/U_{i-1}\Gamma_i$ is a compact abelian group with Lie algebra $\mathfrak{n}_i^*$. Now $\mathfrak{n}_i = \mathfrak{n}_{i-1} + \mathfrak{n}_{\beta_i}$. Choose a base $X_1, \ldots, X_p$ of $\mathfrak{n}_{\beta_i}$ over R and denote the canonical image of X_j in $\mathfrak{n}_i^*$ by X_j^*. Then $X_1^*, \ldots, X_p^*$ is a base of $\mathfrak{n}_i^*$. Put $\mathfrak{v}_j = \mathfrak{n}_{i-1} + \sum_{1 \leq k \leq j} RX_k$ $(o \leq j \leq p)$. Denote by V_j the analytic subgroup of G corresponding to $\mathfrak{v}_j$ and by V_j^* the analytic subgroup of $U_i/U_{i-1}\Gamma_i$ with Lie algebra $\sum_{1 \leq k \leq j} RX_k^*$. Then V_j^* is the image of V_j in $U_i/U_{i-1}\Gamma_i$. Observing that π_{V_j} is defined we put $\varphi_{f,j} = \pi_{V_j} f$ for $f \in \mathcal{L}(\lambda)$, $o \leq j \leq p$, and $\psi_{f,j} = \varphi_{f,j-1} - \varphi_{f,j}$ $(1 \leq j \leq p)$. Then $\pi_{U_{i-1}} f - \pi_{U_i} f = \varphi_{f,o} - \varphi_{f,p} = \sum_{1 \leq j \leq p} \psi_{f,j}$. Using the induction hypothesis we see that Lemma 19 will be proved when we know the following result.

Lemma 20. $\psi_{f,j} \in \mathcal{L}_\alpha(\lambda)$ and $f \longmapsto \psi_{f,j}$ is a continuous mapping of $\mathcal{L}(\lambda)$ into $\mathcal{L}_\alpha(\lambda)$ $(1 \leq j \leq p)$.

Fix j. Write $\varphi_{f,j-1} = \varphi_f$, $V_{j-1} = V$, so $\varphi_f = \pi_V f$. Since $\varphi_f \in C^\infty(G/V\Gamma_\infty)$, the mapping $v^* \longmapsto \varphi_f(xv)$ $(v \in V_j$, $v^* =$ image of v in $V_j^*)$ defines a function belonging to $C^\infty(V_j^*/V_{j-1}^*)$, of which we write the Fourier expansion at $v^* = 1$:

$$\varphi_f(x) = \sum_{q\in Z} \vartheta_{f,q}(x), \qquad \vartheta_{f,q}(x) = \int_0^1 \varphi_f(x \exp tX_j)\, e^{-2\pi iqt}\, dt$$

(where we suppose X_j normalized in appropiate way). We have also

$$\varphi_f(g\iota x) = \sum_{q\in Z} \vartheta_{f,q}(g\iota x) \qquad (g \in \mathcal{G}),$$

and if $q \neq o$:

$$\vartheta_{f,q}(g\iota x) = \int_0^1 \varphi_f(g\iota x \exp tX_j)\, e^{-2\pi iqt}\, dt =$$

$$\frac{1}{(2\pi iq)^r} \int_0^1 \Big(\frac{d^r}{dt^r}\, \varphi_f(g\iota x \exp tX_j)\Big)\, e^{-2\pi iqt}\, dt =$$

$$\frac{1}{(2\pi iq)^r} \int_0^1 \varphi_f(g\iota x \exp tX_j; X_j^r)\, e^{-2\pi iqt}\, dt =$$

$$\frac{1}{(2\pi iq)^r} \int_0^1 \varphi_f(g\, {}^{x}X_j^r\, \iota\, x \operatorname{expt} X_j)\, e^{-2\pi iqt}\, dt$$

for any positive integer r.

Now if $x \in \mathcal{F} = KA_t\omega$ and $x = kap$ $(k \in K,\ a \in A_t,\ p \in \omega)$, then $x = k\, {}^a p\, a = ya$, where y remains in a bounded set Ω. We have then ${}^xX_j = {}^{ya}X_j = e^{\beta_i(\log a)}\, {}^yX_j$. Since $\beta_i \in \Sigma_a$, we have $e^{\beta_i(\log a)} \leq c\, e^{\alpha(\log a)}$ for $a \in A_t$, where c is a constant. And for fixed g and r, $g\, {}^yX_j^r = \sum_k c_k(y) g_k$ with some elements $g_k \in \mathcal{G}$ and bounded functions c_k on Ω. Putting all this together we find

$$\left|\varphi_f(g\, {}^xX_j^r \iota x \exp tX_j)\right| \leq c_o\, e^{r\alpha(H(x))} \sum_k \left|\varphi_f(g_k \iota x \operatorname{expt} X_j)\right| \quad \text{for } x \in \mathcal{F}.$$

Hence

$$\sup_{x\in\mathcal{F}} e^{(\lambda - r\alpha)(H(x))} \left|\vartheta_{f,q}(g\iota x)\right| \leq \frac{c_o}{|2\pi q|^r} \sup_{x\in\mathcal{F}} e^{\lambda(H(x))} \sum_k \int_0^1 \left|\varphi_f(g_k \iota x \operatorname{expt} X_j)\right| dt.$$

Now $\varphi_f = \pi_V f$ and π_V gives a continuous mapping of $\mathcal{L}(\lambda)$ into itself (see §3), so we can choose a continuous seminorm ν on $\mathcal{L}(\lambda)$ such that $c_o \sum_k \nu_{g_k,\lambda}(\varphi_f) \leq \nu(f)$ for all $f \in \mathcal{L}(\lambda)$. From the above inequality we get

$$\nu_{g,\lambda-r\alpha}(\vartheta_{f,q}) \leq \frac{c_o}{|2\pi q|^r} \sum_k \nu_{g_k,\lambda}(\varphi_f) \leq \frac{\nu(f)}{|2\pi q|^r} \qquad \text{for all} \quad q \neq o \ , \quad \text{all} \quad f \in \mathcal{L}(\lambda).$$

We have also

$$\vartheta_{f,o}(x) = \int_{V_j^*/V_{j-1}^*} dv_j^* \int_{V_{j-1}/V_{j-1} \cap \Gamma_i} f(xv_j v_{j-1}) dv_{j-1} = \int_{V_j/V_j \cap \Gamma_i} f(xv_j) dv_j \ ,$$

so $\vartheta_{f,o} = \varphi_{f,j}$. Hence $\psi_{f,j} = \varphi_{f,j-1} - \varphi_{f,j} = \sum_{q \neq o} \vartheta_{f,q}$ and

$$\nu_{g,\lambda-r\alpha}(\psi_{f,j}) \leq \sum_{q \neq o} \nu_{g,\lambda-r\alpha}(\vartheta_{f,q}) \leq \sum_{q \neq o} \frac{1}{|2\pi q|^r} \nu(f) \qquad (f \in \mathcal{L}(\lambda)).$$

Since the seminorms $\nu_{g,\lambda-r\alpha}$ $(g \in \mathcal{G},\ r \in \mathbf{Z},\ r \geq 2)$ determine the topology of $\mathcal{L}_\alpha(\lambda)$, Lemma 20 is proved.

CHAPTER II

§1. Inequalities.

Let (P,A) be any cuspidal pair of G. We denote by $\langle\lambda,\mu\rangle$ the scalar product on $\mathfrak{a}_c^*$ defined by means of the Killing form [if $\lambda \in \mathfrak{a}_c^*$, then $\lambda(H) = B(H_\lambda,H)$ where B is the Killing form of $\mathfrak{g}_c$, $H_\lambda \in \mathfrak{a}_c$; $\langle\lambda,\mu\rangle = B(H_\lambda,H_\mu)$]. Let $\alpha_1,\dots,\alpha_l$ be the simple roots of $(\mathfrak{p},\mathfrak{a})$ and $\lambda_1,\dots,\lambda_l$ elements of $\mathfrak{a}^*$ such that $\langle\lambda_i,\alpha_j\rangle = \delta_{ij}$. Define

$$\mathfrak{a}^+ = \{ H \in \mathfrak{a} : \alpha_i(H) \geq 0 \quad \text{for} \quad 1 \leq i \leq l \},$$
$${}^+\mathfrak{a} = \{ H \in \mathfrak{a} : \lambda_i(H) \geq 0 \quad \text{for} \quad 1 \leq i \leq l \}.$$

If (P,A) is mincuspidal we have $\mathfrak{a}^+ \subset {}^+\mathfrak{a}$ (which is nothing but the well known fact that the inverse of the matrix $(\langle\alpha_i,\alpha_j\rangle)$ has positive entries).

Lemma 21. Assume (P,A) is mincuspidal.

a) Let $\mathfrak{S}$ be a Siegel domain with respect to (P,A) and Ω a compact subset of G. Then

$$\inf_{y_1,y_2\in\Omega,x\in\mathfrak{S}} \lambda_i(H(y_1xy_2)-H(x)) > -\infty \qquad (1 \leq i \leq l).$$

b) We have

$$\inf_{\gamma\in\Gamma} \lambda_i(H(\gamma)) > -\infty \qquad (1 \leq i \leq l).$$

Proof. We may obviously assume that $\underline{G}$ is connected. Fix i $(1 \leq i \leq l)$. There is a multiple $\Lambda = N\lambda_i$ $(N \in Z,\ N \geq 1)$ of λ_i such that there exists an irreducible rational representation (π,V) of $\underline{G}$ defined over $\mathbb{Q}$ with the following property. There is a vector $v \in V_{\mathbb{Q}}$, $v \neq o$, such that $\pi(p)v = v$ for $p \in {}^oP$, $\pi(H)v = \Lambda(H)v$ for $H \in \mathfrak{a}$. We put a Hilbert space structure on V_C such that $\pi(a)$ is self-adjoint for $a \in A$. From $G = KA{}^oP$ one sees that there are constants $c_2 \geq c_1 > o$ such that

$$c_1\, e^{\Lambda(H(x))} \leq |\pi(x)v| \leq c_2\, e^{\Lambda(H(x))} \qquad (x \in G).$$

Let $\Lambda = \Lambda_o,\Lambda_1,\dots,\Lambda_r$ be the weights of π with respect to $\mathfrak{a}$ and let E_i be the orthogonal projection of V on the subspace corresponding to the weight Λ_i. Obviously $H(y_1xy_2) = H(y_1xk_2) + H(y_2)$ if $k_2 \in K$, $k_2^{-1}y_2 \in P$. If $x = kap \in \mathfrak{S} = KA_t\omega$, then $x = k\,{}^apa$ where $k\,{}^ap$ remains bounded. So for $y_1 \in \Omega$, $x \in \mathfrak{S}$, we have $y_1xk_2 = zak_2$

with z in a bounded set. Hence $H(y_1xy_2) - H(x) = H(ak_2) - \log a + R$, where R is bounded $(y_1, y_2 \in \Omega, x \in \mathfrak{S})$. Thus for a) it is sufficient to prove that

$$\inf_{k\in K, a\in A_t} \Lambda(H(ak) - \log a) > -\infty .$$

For $a \in A$, $k \in K$ we have $\pi(ak)v = \sum_{0\leq i\leq r} e^{\Lambda_i(\log a)} E_i\, \pi(k)v$, hence, if $a \in A_t$,

$$|\pi(ak)v|^2 = \sum_{0\leq i\leq r} e^{2\Lambda_i(\log a)} |E_i\pi(k)v|^2 \geq c_o\, e^{2\Lambda(\log a)} |\pi(k)v|^2 \geq c\, e^{2\Lambda(\log a)} ,$$

since $\Lambda - \Lambda_i$ is a linear combination with positive coefficients of the roots $(c > o)$. On the other hand we have seen that $|\pi(ak)v| \leq c_2\, e^{\Lambda(H(ak))}$. This proves a).

In order to prove b) one chooses a lattice L in $V_{\mathbb{Q}}$ such that $\pi(\gamma)L = L$ for $\gamma \in \Gamma$. Then obviously $|\pi(\gamma)v| \geq c_3 > o$ for all $\gamma \in \Gamma$. Hence $e^{\Lambda(H(\gamma))} \geq c_2^{-1} c_3$, which proves b).

<u>Corollary 1</u>. The same inequalities hold when we replace λ_i by any $\lambda \in \mathfrak{a}^*$ such that $\langle\lambda,\alpha\rangle \geq o$ for all $\alpha \in \Sigma^o(P|A)$.

<u>Corollary 2</u>. Let F be a subset of $\Sigma^o(P|A)$. If $\lambda \in \mathfrak{a}_F^*$ such that $\langle\lambda,\alpha\rangle \geq o$ for $\alpha \in \Sigma^o(P_F|A_F)$, then

$$\inf_{y_1, y_2\in\Omega, x\in\mathfrak{S}} \lambda(H_F(y_1xy_2) - H_F(x)) > -\infty$$

and

$$\inf_{\gamma\in\Gamma} \lambda(H_F(\gamma)) > -\infty .$$

<u>Proof</u>. Extend λ to a linear function on $\mathfrak{a}$ by defining it zero on $\mathfrak{a} \cap \mathfrak{m}_F$. Then $\lambda(H(x)) = \lambda(H_F(x))$ $(x \in G)$. If $\alpha \in \Sigma(P|A)$ and if $\bar{\alpha}$ is the restriction of α on $\mathfrak{a}_F$, then $\langle\lambda,\alpha\rangle = \langle\lambda,\bar{\alpha}\rangle \geq o$.

<u>Corollary 3</u>. Let λ be as in Corollary 2. There is a constant c such that

$$\lambda(H_F(y_1xy_2\gamma) - H_F(x)) \geq -c \quad \text{for } y_1, y_2 \in \Omega,\ x \in \mathfrak{S},\ \gamma \in \Gamma .$$

Proof. If $\gamma = kp$, $k \in K$, $p \in P_F$, then $H_F(y_1xy_2\gamma) - H_F(x) = H_F(y_1xy_2k) - H_F(x) + H_F(\gamma)$. Apply Corollary 2 with ΩK instead of Ω .

Changing the notation we suppose now that (P,A) is a cuspidal pair and $(P_o,A_o) \prec (P,A)$ a mincuspidal pair. In the following lemma $H = H_{P|A}$.

Lemma 22. Let $\mathfrak{F}$ be a Siegel domain with respect to (P_o,A_o) and Ω a compact subset of G . We can choose an element $H_o \in \mathfrak{a}$ such that

$$H(y_1xy_2\gamma) - H(x) \in {}^+\mathfrak{a} + H_o \quad \text{for all} \quad y_1,y_2 \in \Omega,\ x \in \mathfrak{F},\ \gamma \in \Gamma.$$

Proof. This follows immediately from Corollary 3 of Lemma 21.

Definition. $(\mathfrak{a}^*)^- = \{\lambda \in \mathfrak{a}^* : \langle\lambda + \varrho,\alpha\rangle < o \text{ for } \alpha \in \Sigma^o(P|A)\}$,

$(\mathfrak{a}_c^*)^- = \{\Lambda \in \mathfrak{a}_c^* : \mathrm{Re}(\Lambda) \in (\mathfrak{a}^*)^-\}$.

§2. Eisenstein series.

If $P = MAU$ is a cuspidal subgroup of G , σ a finite-dimensional representation of K and φ a σ_M-function on M (see Chap. I,§2), then φ can be extended to a σ-function on G by the formula

$$\varphi(k\ mau) = \sigma(k)\varphi(m) \qquad (k \in K,\ m \in M,\ a \in A,\ u \in U).$$

Although a decomposition $x = k\ mau$ is not unique, the above formula defines φ without ambiguity on G . When we speak of σ_M-functions on M , it will always be understood that they are extended in this way to σ-functions on G .

Let (P,A) be a cuspidal pair, $P = MAU$. Let (P_o,A_o) be a mincuspidal pair and $\mathfrak{F}_o$ a Siegel domain with respect to it. Fix $\xi \in G_{\mathbb{Q}}$ such that $(P',A') = {}^{\xi}(P,A) \succ (P_o,A_o)$. For $\varphi \in L^1(M/\Gamma_M,\sigma_M)$, $\lambda \in \mathfrak{a}_c^*$, put $\varphi_\lambda(x) = \varphi(x)\ e^{(\lambda-\varrho)(H(x))}$, where ϱ is, as usual, one half of the sum of the positive roots of (P,A). Then φ_λ is a σ-function and is right-invariant under $U(\Gamma \cap P)$.

Lemma 23. There is a number N such that for any compact set Ω in G there exist a number c and an element H_1 of $\mathfrak{a}$ such that

$$\sum_{\gamma\in\Gamma/\Gamma\cap P} \int_{\Omega} \left|\varphi_\lambda(y^{-1}xy_o\gamma)\right| dy \leq c \, \|\varphi\|_1 \, \|x\|^N \, \frac{e^{(\lambda'+\varrho')(H'(x)) + (\lambda+\varrho)(H_1)}}{\prod_{\alpha\in\Sigma^o(P|A)} |\langle\lambda+\varrho,\alpha\rangle|}$$

for $\varphi \in L^1(M/\Gamma_M,\sigma_M)$, $y_o \in \Omega$, $x \in \mathfrak{S}_o$, $\lambda \in (\mathfrak{a}^*)^-$ [Here $H' = H_{P'|A'}$ and $\lambda' = {}^{\xi}\lambda$].

<u>Proof</u>. If β is the characteristic function of the compact set Ω , then

$$\int_{\Omega} |\varphi_\lambda(y^{-1}x)| dy = \int_G \beta(xy^{-1}) |\varphi_\lambda(y)| dy = \int_{G/\Gamma\cap P} \sum_{\Gamma\cap P} \beta(x\gamma y^{-1}) |\varphi_\lambda(y)| dy \, ,$$

$$\sum_{\Gamma/\Gamma\cap P} \int_{\Omega} |\varphi_\lambda(y^{-1}x\gamma)| dy = \int_{G/\Gamma\cap P} \sum_{\Gamma} \beta(x\gamma y^{-1}) |\varphi_\lambda(y)| dy \, .$$

Hence, by Lemma 9,

$$\sum_{\Gamma/\Gamma\cap P} \int_{\Omega} |\varphi_\lambda(y^{-1}x\gamma)| dy \leq c_1 \, \|x\|^N \int_{\Omega^{-1}x\Gamma/\Gamma\cap P} |\varphi_\lambda(y)| dy \qquad (x \in G),$$

where N is independent of Ω . If $\mathfrak{S}_o = KA_{o,t}\omega_o$, then $K\,{}^{\xi-1}(A_{o,t}\omega_o) = \mathfrak{S}$ is a Siegel domain with respect to the mincuspidal pair ${}^{\xi-1}(P_o,A_o)$ which is dominated by (P,A). Put $\Omega_1 = \xi^{-1}\Omega \cup \Omega^{-1}K\xi$ and choose $H_1 \in \mathfrak{a}$ such that $H(y_1xy_2\gamma) - H(x) \in {}^+\mathfrak{a} + H_1$ for $y_1,y_2 \in \Omega_1$, $x \in \mathfrak{S}$, $\gamma \in \Gamma$ (Lemma 22). Let $x_o \in \mathfrak{S}_o$. Writing $x_o = k_o\,{}^{\xi}p$ with $k_o \in K$, $p \in \mathfrak{S} \cap {}^{\xi-1}P_o$ we have $y^{-1}x_oy_o\gamma = (y^{-1}k_o\xi)p(\xi^{-1}y_o)\gamma$, so $H(y^{-1}x_oy_o\gamma) - H(p) \in {}^+\mathfrak{a} + H_1$ for $y_o,y \in \Omega$, $\gamma \in \Gamma$. Since $H(p) = {}^{\xi-1}H'(x_o)$, this gives $\Omega^{-1}x_o\Omega\Gamma \subset K\,{}^{\xi-1}(a'(x_o))a_1\,{}^+A\,{}^oP = B$ (where ${}^+A = \exp{}^+\mathfrak{a}$, $a_1 = \exp H_1$). So, for $x_o \in \mathfrak{S}_o$, $y_o \in \Omega$ we have

$$\sum_{\Gamma/\Gamma\cap P} \int_{\Omega} |\varphi_\lambda(y^{-1}x_oy_o\gamma)| dy \leq c_1 \, \|x_oy_o\|^N \int_{B/\Gamma\cap P} |\varphi_\lambda(y)| dy \, .$$

Using the integration formula

$$\int_G f(y)\,dy = \int_{K} \int_{A} \int_{M} \int_{U} f(k\,amu)\, e^{2\varrho(\log a)}\, dk\, da\, dm\, du$$

we get

$$\int_{B/\Gamma\cap P} |\varphi_\Lambda(y)|\,dy = \|\varphi\|_1 \int_{\xi^{-1}(a'(x_o))a_1{}^+A} e^{(\lambda+\varrho)(\log a)}\,da =$$

$$\|\varphi\|_1\, e^{(\lambda'+\varrho')(H'(x_o)) + (\lambda+\varrho)(H_1)} \int_{^+A} e^{(\lambda+\varrho)(\log a)}\,da\,.$$

If α_i and λ_i $(1 \leq i \leq l)$ are as in the beginning of §1, then $\lambda+\varrho = \sum_{1\leq i\leq l} \langle\lambda+\varrho,\alpha_i\rangle\lambda_i$ and

$$\int_{^+A} e^{(\lambda+\varrho)(\log a)}\,da = \prod_{1\leq i\leq l} \int_o^\infty e^{\langle\lambda+\varrho,\alpha_i\rangle\lambda_i}\,d\lambda_i = \prod_{1\leq i\leq l} \frac{-1}{\langle\lambda+\varrho,\alpha_i\rangle}$$

if $\langle\lambda+\varrho,\alpha_i\rangle < o$ for all i. Since $c_1\|x_oy_o\|^N \leq c\|x_o\|^N$ we get the wanted inequality.

Corollary 1. Let Ω and ω^* be compact sets in G and $(\mathfrak{a}^*)^-$ respectively. For any $\beta \in C_c(G)$ and $\varphi \in L^1(M/\Gamma_M,\sigma_M)$ the series

$$\sum_{\Gamma/\Gamma\cap P} |(\beta*\varphi_\Lambda)(x\gamma)|$$

converges uniformly for $x \in \Omega$ and $\mathrm{Re}(\Lambda) \in \omega^*$.

Proof. Assume that $\Omega \supset \mathrm{Supp}(\beta)$. Then

$$|\beta*\varphi_\Lambda(x\gamma)| = |\int_G \beta(y)\varphi_\Lambda(y^{-1}x\gamma)\,dy| \leq \|\beta\|_\infty \int_\Omega |\varphi_\Lambda(y^{-1}x\gamma)|\,dy \leq$$

$$\|\beta\|_\infty \int_{\Omega^{-1}\Omega} |\varphi_\Lambda(y^{-1}\gamma)|\,dy \qquad \text{for } x \in \Omega,\ \gamma \in \Gamma\ ;$$

$$|\varphi_\Lambda(z)| = |\varphi(z)|\, e^{(\mathrm{Re}(\Lambda)-\varrho)(H(z))} \leq |\varphi(z)| \sup_{\lambda\in\omega^*} e^{(\lambda-\varrho)(H(z))}\ , \text{ etc.}$$

Corollary 2. For any $\beta \in C_c^\infty(G)$ and $\varphi \in L^1(M/\Gamma_M,\sigma_M)$ the series

$$\sum_{\Gamma/\Gamma\cap P} (\beta*\varphi_\Lambda)(x\gamma)$$

converges absolutely for $x \in G$, $\Lambda \in (\mathfrak{a}_c^*)^-$. Its sum is a C^∞-function on $G \times (\mathfrak{a}_c^*)^-$ which is holomorphic in Λ. The series can be differentiated term by term with respect to (x,Λ).

Proof. This is an obvious consequence of Corollary 1.

Corollary 3. Let Ω and ω^* be compact subsets of G and $(\mathfrak{a}^*)^-$ respectively. For any $\varphi \in L^1(M/\Gamma_M,\sigma_M,\chi)$ (where χ is a representation of $\mathfrak{Z}_M$ on the space V of σ commuting with σ_M) and any $g \in \mathfrak{G}$ the series

$$\sum_{\Gamma/\Gamma\cap P} |(g*\varphi_\Lambda)(x\gamma)|$$

converges uniformly with respect to (x,Λ) on $\Omega \times (\omega^* + i\mathfrak{a}^*)$.

Proof. Let $\mathfrak{Z}_1 = \mathfrak{Z}_M\mathfrak{A}$ denote, as in Chap. I, §6, the center of the universal enveloping algebra of $\mathfrak{m} + \mathfrak{a}$. Fix $\varphi \in L^1(M/\Gamma_M,\sigma_M,\chi)$. Since φ_Λ is right-invariant under U, $X\varphi_\Lambda = 0$ for $X \in \mathfrak{n}$, so $z\varphi_\Lambda = \mu(z)'\varphi_\Lambda$ for $z \in \mathfrak{Z}$, where μ is as defined l.c. Now $\mu(z) = \sum \zeta_i q_i$ with certain $\zeta_i \in \mathfrak{Z}_M$, $q_i \in \mathfrak{A} = S(\mathfrak{a}_c)$. Put $\mu_\Lambda(z) = \sum q_i(\Lambda)\zeta_i$. Then (in very careless notation)

$$z\varphi_\Lambda = \mu(z)'\varphi_\Lambda = e^{-\varrho}\mu(z)e^{\Lambda}\varphi = e^{\Lambda-\varrho}e^{-\Lambda}\mu(z)e^{\Lambda}\varphi = e^{\Lambda-\varrho}\sum q_i(\Lambda)\zeta_i\varphi =$$

$$= e^{\Lambda-\varrho}\mu_\Lambda(z)\varphi = (\mu_\Lambda(z)\varphi)_\Lambda = (\varphi\chi(\mu_\Lambda(z)))_\Lambda = \varphi_\Lambda\chi(\mu_\Lambda(z)).$$

In particular φ_Λ is a $\mathfrak{Z}$-finite function on G. Since it is also K-finite, we can choose a function $\alpha \in C_c^\infty(G)$ such that $\varphi_\Lambda = \alpha*\varphi_\Lambda$. Corollary 3 is now a particular case of Corollary 1.

Definition. $E(\Lambda:\varphi:x) = \sum_{\Gamma/\Gamma\cap P} \varphi_\Lambda(x\gamma)$ $(\varphi \in L^2(M/\Gamma_M,\sigma_M,\chi),\ \Lambda \in (\mathfrak{a}_c^*)^-,\ x \in G)$.

By Corollary 2 of Lemma 23, $E(\Lambda:\varphi:x)$ is a C^∞-function on $G \times (\mathfrak{a}_c^*)^-$ which is holomorphic in Λ. As a function of x it is right-invariant under Γ, it is a σ-function and

$$z\,E(\Lambda:\varphi) = E(\Lambda:\varphi)\chi(\mu_\Lambda(z)) \qquad (z \in \mathfrak{Z})$$

[See the proof of Cor. 3 of Lemma 23].

Let us return to the situation of Lemma 23. Assume that $\varphi \in L^2(M/\Gamma_M,\sigma_M,\chi)$ for some χ. Choose $\beta \in C_c^\infty(G)$ such that $\beta*\varphi_\Lambda = \varphi_\Lambda$, where Λ is a fixed element of $(\mathfrak{a}_c^*)^-$. Then we have, for $x \in \mathfrak{F}_0$, $y_0 \in \Omega$,

$$|\varphi_\Lambda(xy_o\gamma)| = |\int_G \beta(y)\varphi_\Lambda(y^{-1}xy_o\gamma)dy| \leq \|\beta\|_\infty \int_{Supp(\beta)} |\varphi_\Lambda(y^{-1}xy_o\gamma)|dy ,$$

so that we can apply Lemma 23 in order to estimate $\sum_{\Gamma/\Gamma\cap P} |\varphi_\Lambda(xy_o\gamma)|$. However, the function β depends on Λ, which makes it necessary to proceed more carefully.

Let $I_c^\infty(G)$ denote the space of all functions $\beta \in C_c^\infty(G)$ such that $\beta(kxk^{-1}) = \beta(x)$ $(x \in G, k \in K)$. For any $\Lambda \in \mathfrak{a}_c^*$ we have a representation π_Λ of $I_c^\infty(G)$ on $L = L^2(M/\Gamma_M,\sigma_M,\chi)$ defined by

$$\beta*\varphi_\Lambda = (\pi_\Lambda(\beta)\varphi)_\Lambda \qquad (\varphi \in L).$$

In fact, an easy calculation shows that $\beta*\varphi_\Lambda = (\hat\beta_\Lambda*\varphi)_\Lambda$ where

$$\hat\beta_\Lambda(m) = \iiint_{K\times A\times U} \beta(k^{-1}mau)\ e^{-(\Lambda-\varrho)(\log a)}\ \sigma(k)dk\ da\ du \quad (m \in M).$$

The mapping $\Lambda \longmapsto \pi_\Lambda(\beta)$ is an entire function $\mathfrak{a}_c^* \to End(L)$. We know that L has finite dimension $(L \subset \mathcal{A}(M/\Gamma_M,\sigma_M,\chi)$ because $\alpha \in C_c^\infty(M)$ and $f \in L^2(M/\Gamma_M)$ imply $\alpha*f \in \mathcal{A}^\infty(M/\Gamma_M)$: see the first part of the proof of Lemma 13).

Fix $\Lambda_o \in (\mathfrak{a}_c^*)^-$. Choose $\beta \in I_c^\infty(G)$ such that $\beta*\varphi_{\Lambda_o} = \varphi_{\Lambda_o}$ for <u>all</u> $\varphi \in L$. Now $\pi_{\Lambda_o}(\beta) = 1$, so there is a compact neighbourhood ω^* of Λ_o in $(\mathfrak{a}_c^*)^-$ such that $\det(\pi_\Lambda(\beta)) \neq 0$ for $\Lambda \in \omega^*$. Let $\varphi_1,\dots,\varphi_N$ be a base for L and put $\pi_\Lambda(\beta)^{-1}\varphi = \sum t_i(\Lambda)\varphi_i$ $(\Lambda \in \omega^*)$ for a fixed $\varphi \in L$. Then $\varphi = \sum t_i(\Lambda)\pi_\Lambda(\beta)\varphi_i$ and $\varphi_\Lambda = \sum t_i(\Lambda)\beta*\varphi_{i,\Lambda}$. The functions t_i are of course bounded on ω^*. Now it is clear that Lemma 23 gives the following estimate.

<u>Lemma 24</u>. The notations being as in Lemma 23, let φ be a function in $L^2(M/\Gamma_M,\sigma_M,\chi)$ for some χ. Given a compact set Ω in G there exists a locally bounded function c on $(\mathfrak{a}^*)^-$ such that

$$\sum_{\Gamma/\Gamma\cap P} |\varphi_\Lambda(xy\gamma)| \leq c(\Lambda)\|x\|^N e^{(\lambda'+\varrho')(H'(x))}$$

for $x \in \mathcal{G}_o$, $y \in \Omega$, $\Lambda \in (\mathfrak{a}_c^*)^-$, $\lambda = Re(\Lambda)$.

<u>Corollary</u>. If $\varphi \in L^2(M/\Gamma_M,\sigma_M,\chi)$ and $\Lambda \in (\mathfrak{a}_c^*)^-$, then $E(\Lambda:\varphi) \in \mathcal{A}(G/\Gamma,\sigma)$.

<u>Remarks</u>. 1) We have in particular (take $\varphi = 1$) proved Godement's criterium: If Ω and ω^* are compact subsets of G and $(\mathfrak{a}^*)^-$ respectively, then the series

$$\sum_{\Gamma/\Gamma\cap P} e^{(\lambda-\varrho)(H(x\gamma))}$$

converges uniformly for $\lambda \in \omega^*$, $x \in \Omega$.

2) Take in Lemma 23 $\varphi = 1$. There is a constant c_λ such that $\varphi_\lambda(x) \leq c_\lambda \varphi_\lambda(y^{-1}x)$ for all $y \in \Omega$, $x \in G$, hence $\mathrm{vol}(\Omega)\varphi_\lambda(x) \leq c_\lambda \int_\Omega \varphi_\lambda(y^{-1}x)dy$. Moreover

$$c_\lambda \leq \sup_{y\in\Omega, k\in K} e^{-(\lambda-\varrho)(H(y^{-1}k))} \leq c_o\, e^{(\lambda-\varrho)(H_o)} \quad \text{for all } \lambda \text{ with constant } c_o$$

and H_o . So we get finally

$$\sum_{\Gamma/\Gamma\cap P} e^{(\lambda-\varrho)(H(x\gamma))} \leq c_1 \frac{e^{(\lambda-\varrho)(H_1)}}{\prod_{\alpha\in\Sigma^o} |\langle\lambda+\varrho,\alpha\rangle|}$$

for all $\lambda \in (\mathfrak{a}^*)^-$, $x \in \Omega$ (c_1 and H_1 depending on Ω).

§3. <u>The functions E_f.</u>

Let $P = MAU$ be a cuspidal subgroup of G , and σ and χ representations of K resp. $\mathfrak{Z}_M$ on a space V as usual. We denote by $\mathcal{D}_{P|A}(\sigma,\chi)$ the subspace of $C^\infty(G/U(\Gamma\cap P),\sigma)$ spanned by the functions f of the form $f(x) = \varphi(x)v(a(x))$ with $\varphi \in L^2(M/\Gamma_M,\sigma_M,\chi)$, $v \in C_c^\infty(A)$, and by $\dot{\mathcal{D}}_{P|A}(\sigma,\chi)$ the subspace defined in the same way but with φ restricted to ${}^oL^2(M/\Gamma_M,\sigma_M,\chi)$.

For any $f \in \mathcal{D}_{P|A}(\sigma,\chi)$ we define a Fourier transform by

$$\hat{f}(\Lambda:x) = \int_A f(xa)\, e^{-(\Lambda-\varrho)(H(xa))}\, da \quad (\Lambda \in \mathfrak{a}_c^*,\ x \in G).$$

If f is of the form $f(x) = \varphi(x)v(a(x))$, then $\hat{f}(\Lambda:x) = \varphi(x)\hat{v}(\Lambda)$, where

$$\hat{v}(\Lambda) = \int_A v(a)\, e^{-(\Lambda-\varrho)(\log a)}\, da .$$

The inverse transformation is given by

$$f(x) = \int_{\mathrm{Re}(\Lambda)=\lambda} \hat{f}(\Lambda:x)\, e^{(\Lambda-\varrho)(H(x))}\, d\Lambda_I \qquad (x \in G)$$

where λ is arbitrary in $\mathfrak{a}^*$ $(\Lambda_I = \mathrm{Im}(\Lambda))$. Clearly $\hat{f}(\Lambda) \in L^2(M/\Gamma_M,\sigma_M,\chi)$ and the mapping $\Lambda \longmapsto \hat{f}(\Lambda)$ is holomorphic on $\mathfrak{a}_c^*$.

Lemma 25. Let $f \in \mathcal{D}_{P|A}(\sigma,\chi)$. The series $\sum_{\Gamma/\Gamma\cap P} |f(x\gamma)|$ converges uniformly on every compact subset of G and its sum lies in $\mathcal{A}(G/\Gamma,\sigma)$.

Proof. It is sufficient to consider a function $f(x) = \varphi(x)v(a(x))$. Choose any $\lambda \in (\mathfrak{a}^*)^-$. If $c = \sup_{a\in A}|v(a)|e^{-(\lambda-\varrho)(\log a)}$, then $|f(x)| \leq c|\varphi(x)|\, e^{(\lambda-\varrho)(H(x))}$. So our statements follow from the results of §2.

Definition. $E_f(x) = \sum_{\Gamma/\Gamma\cap P} f(x\gamma) \qquad (f \in \mathcal{D}_{P|A}(\sigma,\chi),\ x \in G)$.

Remark. This is in fact a finite sum.

By Lemma 25, $E_f \in \mathcal{A}(G/\Gamma,\sigma)$.

Lemma 26. Let $f \in {}^0\mathcal{D}_{P|A}(\sigma,\chi)$ and $g \in \mathcal{A}(G/\Gamma)$. Put $E_f^*(x) = \sum_{\Gamma/\Gamma\cap P} |f(x\gamma)|$. Then $\int_{G/\Gamma} E_f^*(x)\, |g(x)|\, dx < \infty$.

Proof. Observing that E_f^* is left-invariant under K (σ is unitary) we may assume that g is left-invariant under K and positive. Moreover, assume that $f(x) = \varphi(x)v(a(x))$ with $\varphi \in {}^0L^2(M/\Gamma_M,\sigma_M,\chi)$, $v \in C_c^\infty(A)$.

$$\int_{G/\Gamma} E_f^*(x)g(x)dx = \int_{G/\Gamma\cap P} |f(x)|g(x)dx = \int_A e^{2\varrho(\log a)}|v(a)|da \int_{M/\Gamma_M} |\varphi(m)|dm \int_{U/U\cap\Gamma} g(amu)du.$$

If $g(x) \leq c\|x\|^r$ $(x \in G)$, then $\int_{U/U\cap\Gamma} g(amu)du \leq c_1\|a\|^r\|m\|^r$. The remaining integrals $\int_A e^{2\varrho(\log a)}|v(a)|\,\|a\|^r\, da$ and $\int_{M/\Gamma_M} |\varphi(m)|\,\|m\|^r\, dm$ are finite: for the first one this is trivial and for the second one it follows from Lemmas 12 and 5.

Corollary. $E_f^* \in L^2(G/\Gamma)$ and $E_f \in L^2(G/\Gamma,\sigma)$ if $f \in {}^0\mathcal{D}_{P|A}(\sigma,\chi)$.

Lemma 27. For $f \in {}^0\mathcal{D}_{P|A}(\sigma,\chi)$, $g \in \mathcal{A}(G/\Gamma,\sigma)$ we have

$$\int_{G/\Gamma} (E_f(x), g(x))dx = \int_A e^{2\varrho(\log a)} da \int_{M/\Gamma_M} (f(am), g^o(am))dm$$

where g^o is defined by

$$g^o(x) = \int_{U/U\cap\Gamma} g(xu)du.$$

Proof. Trivial, the integral over G/Γ being absolutely convergent by Lemma 26.

Lemma 28. If $f \in \mathfrak{D}_{P|A}(\sigma,\chi)$ then

$$E_f(x) = \int_{\mathrm{Re}(\Lambda)=\lambda} E(\Lambda:\hat{f}(\Lambda):x)\, d\Lambda_I \qquad (x \in G)$$

where λ is arbitrary in $(\mathfrak{a}^*)^-$.

Proof. We have only to verify that in

$$\sum_{\Gamma/\Gamma\cap P} \int_{\mathrm{Re}(\Lambda)=\lambda} \hat{f}(\Lambda:x\gamma)\, e^{(\Lambda-\varrho)(H(x\gamma))}\, d\Lambda_I$$

summation and integration may be interchanged if $\lambda \in (\mathfrak{a}^*)^-$. Assume f is of the form $f(x) = \varphi(x)v(a(x))$. Then

$$\sum_{\Gamma/\Gamma\cap P} \int_{\mathrm{Re}(\Lambda)=\lambda} |\hat{f}(\Lambda:x\gamma)\, e^{(\Lambda-\varrho)(H(x\gamma))}|\, |d\Lambda| =$$

$$\sum_{\Gamma/\Gamma\cap P} |\varphi_\Lambda(x\gamma)| \int_{\mathfrak{a}^*} |\hat{v}(\lambda+i\mu)|\, d\mu < \infty$$

because of the results of §2 and the fact that $\mu \longmapsto \hat{v}(\lambda+i\mu)$, as Fourier transform of a function in $C_c^\infty(A)$, is a Schwartz function, so is certainly integrable.

§4. The constant term of the Fourier series of E_Λ.

We need first some algebraic preparation.

Let (P_1,A_1) and (P_2,A_2) be two cuspidal pairs. Suppose there exists $x \in G$ such that ${}^xA_1 = A_2$. The homomorphism $\underline{\underline{A}}_1 \to \underline{\underline{A}}_2$ defined by x is then defined over $\mathbb{Q}$ and from this one sees easily that ${}^x\underline{\underline{P}}_1$ is defined over $\mathbb{Q}$. So ${}^xP_1 = {}^\xi P_1$ for some $\xi \in G_{\mathbb{Q}}$ and since ${}^\xi A_1$ and A_2 are two split components of ${}^\xi P_1$, we may assume (after multiplying ξ with an element of $(P_1)_{\mathbb{Q}}$) that ${}^\xi A_1 = A_2$. Then $\xi^{-1}x$ is contained in P_1 and normalizes A_1, so it centralizes A_1. So the homomorphism $A_1 \to A_2$ defined by x can also be defined by an element of $G_{\mathbb{Q}}$.

<u>Definition</u>. Two cuspidal subgroups P_1 and P_2 of G are called <u>associate</u> if for some split components A_1, A_2 of P_1, P_2 there exists an element y of $G_{\mathbb{Q}}$ such that ${}^{y}A_1 = A_2$.

If this condition is fulfilled for one choice of A_1, A_2, it is fulfilled for any choice of A_1, A_2, since two split components of the same cuspidal group are conjugate under $G_{\mathbb{Q}}$.

For any two cuspidal pairs (P_1, A_1), (P_2, A_2) of G we denote by $w(\mathfrak{a}_1, \mathfrak{a}_2)$ the set of all mappings $\mathfrak{a}_1 \to \mathfrak{a}_2$ of the form $\mathrm{Ad}(y)|_{\mathfrak{a}_1}$ with $y \in G_{\mathbb{Q}}$ such that ${}^{y}\mathfrak{a}_1 = \mathfrak{a}_2$. Then $w(\mathfrak{a}_1, \mathfrak{a}_2)$ is not empty if and only if P_1 and P_2 are associate. If P_1, P_2, P_3 are associate, then $w(\mathfrak{a}_1, \mathfrak{a}_3) = w(\mathfrak{a}_2, \mathfrak{a}_3) w(\mathfrak{a}_1, \mathfrak{a}_2)$ and $w(\mathfrak{a}_2, \mathfrak{a}_1) = w(\mathfrak{a}_1, \mathfrak{a}_2)^{-1}$. In particular put $w(\mathfrak{a}) = w(\mathfrak{a}, \mathfrak{a})$; $w(\mathfrak{a})$ is a group, isomorphic to $N(A)_{\mathbb{Q}}/Z(A)_{\mathbb{Q}}$, hence finite. If $s \in w(\mathfrak{a}_1, \mathfrak{a}_2)$, then $w(\mathfrak{a}_1, \mathfrak{a}_2) = s w(\mathfrak{a}_1) = w(\mathfrak{a}_2) s$; $w(\mathfrak{a}_1, \mathfrak{a}_2)$ is finite. We recall

<u>Bruhat's lemma</u>. Let (P, A) be mincuspidal, $w(\mathfrak{a}) = N(A)_{\mathbb{Q}}/Z(A)_{\mathbb{Q}}$. For each $s \in w(\mathfrak{a})$, fix $w_s \in N(A)_{\mathbb{Q}}$ such that $\mathrm{Ad}(w_s) = s$ on $\mathfrak{a}$. Then $G_{\mathbb{Q}} = \bigcup_{s \in w(\mathfrak{a})} U_{\mathbb{Q}} w_s P_{\mathbb{Q}}$ and this union is disjoint.

<u>Lemma 29</u>. Let $P_i = M_i A_i U_i$ $(i = 1, 2)$ be two cuspidal subgroups of G. Put $\pi_i = \pi_{P_i | M_i}$. Then

a) ${}^{*}P_1 = \pi_1({}^{o}P_1 \cap P_2)$ is a cuspidal subgroup of M_1 whose unipotent radical is $\pi_1({}^{o}P_1 \cap U_2)$.

b) Suppose ${}^{*}P_1 = M_1$. Then there exists $\xi \in G_{\mathbb{Q}}$ such that $A_1 \supset {}^{\xi}A_2$.

<u>Proof</u>. Fix a mincuspidal pair $(P, A) \prec (P_1, A_1)$. Choose $\xi \in G_{\mathbb{Q}}$ such that $(P_2, A_2) \succ {}^{\xi}(P, A)$. Then $(P_1, A_1) = (P_{F_1}, A_{F_1})$ and $(P_2, A_2) = {}^{\xi}(P_{F_2}, A_{F_2})$ for some $F_1, F_2 \subset \Sigma^{o}(P|A)$. Write $\xi = u_o w p_o$, $u_o = u_1 u_2$ with $u_o \in U_{\mathbb{Q}}$, $w \in N(A)_{\mathbb{Q}}$, $p_o \in P_{\mathbb{Q}}$, $u_1 \in (M_1 \cap U)_{\mathbb{Q}}$, $u_2 \in (U_1)_{\mathbb{Q}}$. Then ${}^{o}P_1 \cap P_2 = {}^{o}P_{F_1} \cap {}^{\xi}P_{F_2} = {}^{u_o}({}^{o}P_{F_1} \cap {}^{w}P_{F_2})$ and

$\pi({}^{o}P_1 \cap P_2) = {}^{u_1}(\pi({}^{o}P_{F_1} \cap {}^{w}P_{F_2}))$ where $\pi = \pi_1$. Now consider first the Lie algebra $\pi({}^{o}\mathfrak{g}_{F_1} \cap {}^{w}\mathfrak{g}_{F_2}) = \mathfrak{m}_{F_1} \cap ({}^{w}\mathfrak{g}_{F_2} + \mathfrak{n}_{F_1})$. We claim that this is equal to $\mathfrak{m}_{F_1} \cap {}^{w}\mathfrak{g}_{F_2}$. In order to prove this, consider the decomposition $\mathfrak{g} = \sum \mathfrak{g}_{\alpha}$ of $\mathfrak{g}$ as sum of root spaces with respect to $\mathfrak{a}$; denote by σ_α the projection of $\mathfrak{g}$ on $\mathfrak{g}_\alpha$. Then $\mathfrak{m}_{F_1}$ and ${}^{w}\mathfrak{g}_{F_2}$ are stable under each σ_α. Let Σ_1 be the set of all roots of $(\mathfrak{g}, \mathfrak{a})$ which vanish on $\mathfrak{a}_{F_1}$ and put $\sigma_1 = \sigma_o + \sum_{\alpha\in\Sigma_1} \sigma_\alpha$. If $X = Y + Z \in \mathfrak{m}_{F_1}$, $Y \in {}^{w}\mathfrak{g}_{F_2}$, $Z \in \mathfrak{n}_{F_1}$, then $X = \sigma_1 X = \sigma_1 Y + \sigma_1 Z = \sigma_1 Y \in {}^{w}\mathfrak{g}_{F_2}$. So the Lie algebra of $\pi({}^{o}P_{F_1} \cap {}^{w}P_{F_2})$ is $\mathfrak{m}_{F_1} \cap {}^{w}\mathfrak{g}_{F_2}$. Clearly this Lie algebra is parabolic in $\mathfrak{m}_{F_1}$ and is a $\mathbb{Q}$-space. We have still to prove that $\pi({}^{o}P_{F_1} \cap {}^{w}P_{F_2})$ is the full normalizer of $\mathfrak{m}_{F_1} \cap {}^{w}\mathfrak{g}_{F_2}$ in M_1. A way of doing this is to show that the complex group $\pi({}^{o}\underline{\underline{P}}_{F_1} \cap {}^{w}\underline{\underline{P}}_{F_2}) = \underline{\underline{Q}}$ meets every component of $\underline{\underline{M}}_1$. We know that $(P \cap M_1, A \cap M_1)$ is mincuspidal in M_1 (Lemma 2), the corresponding group "M" being actually M. This implies that $\underline{\underline{M}}$ meets every component of $\underline{\underline{M}}_1$, and so does $\underline{\underline{Q}}$, because $\underline{\underline{Q}} \supset \underline{\underline{M}}$. We have now proved that $\pi({}^{o}P_{F_1} \cap {}^{w}P_{F_2})$ is a cuspidal subgroup of M_1. Then $\pi({}^{o}P_1 \cap P_2)$ is one too. The proof that its unipotent radical is $\pi({}^{o}P_1 \cap U_2)$ is easy.

Assume that ${}^{*}P_1 = M_1$. Then $\pi({}^{o}P_{F_1} \cap {}^{w}P_{F_2}) = M_{F_1}$ and $\mathfrak{m}_{F_1} \subset {}^{w}\mathfrak{g}_{F_2}$. So $\mathfrak{a}_{F_1} + \mathfrak{m}_{F_1} \subset {}^{w}\mathfrak{g}_{F_2}$ and, in fact, one sees easily that $\mathfrak{a}_{F_1} + \mathfrak{m}_{F_1} \subset {}^{w}(\mathfrak{a}_{F_2} + \mathfrak{m}_{F_2})$. So $\mathfrak{z}(\mathfrak{a}_{F_1}) \subset \mathfrak{z}({}^{w}\mathfrak{a}_{F_2})$, hence ${}^{w}\mathfrak{a}_{F_2}$ centralizes $\mathfrak{z}(\mathfrak{a}_{F_1})$ and is contained in $\mathfrak{a}$, so ${}^{w}\mathfrak{a}_{F_2} \subset \mathfrak{a}_{F_1}$ and $A_1 \supset {}^{w\xi^{-1}}A_2$.

Lemma 30. The notations being as in Lemma 29 we have:

a) $P_1 \cap U_2 = {}^{o}P_1 \cap U_2$.

b) If ${}^{*}U_1$ is the unipotent radical of ${}^{*}P_1$, then ${}^{*}U_1/{}^{*}U_1 \cap \pi_1(P_1 \cap \Gamma \cap P_2)$ is compact.

Proof. a) is trivial. Since ${}^{o}P_1 \cap U_2/{}^{o}P_1 \cap U_2 \cap \Gamma$ is compact, ${}^{*}U_1/\pi_1({}^{o}P_1 \cap U_2 \cap \Gamma)$ is compact; $\pi_1({}^{o}P_1 \cap U_2 \cap \Gamma) \subset {}^{*}U_1 \cap \pi_1(P_1 \cap \Gamma \cap P_2)$. This gives b).

In the remainder of this paragraph we fix two cuspidal pairs (P_i, A_i) $(i = 1,2)$ with decompositions $P_i = M_i A_i U_i$, and a finite-dimensional unitary representation σ of K on a space V. For abbreviation we write $\sigma_i = \sigma_{M_i}$, $\Gamma_i = \Gamma_{M_i}$. For

$\varphi \in L^2(M_1/\Gamma_1,\sigma_1,\chi_1)$ (where χ_1 is a representation of $\mathfrak{Z}_{M_1}$ on V) and $\Lambda \in (\mathfrak{a}^*_{1,c})^-$ we consider the Eisenstein series

$$E_\Lambda(x) = \sum_{\Gamma/\Gamma\cap P_1} \varphi_\Lambda(x\gamma)$$

and its 0-th Fourier coefficient with respect to U_2 :

$$E^o_\Lambda(x) = \int_{U_2/U_2\cap\Gamma} E_\Lambda(xu)\,du \ .$$

Here the measure du is such that $\int_{U_2/U_2\cap\Gamma} du = 1$. For any $f \in \mathcal{A}(G/\Gamma,\sigma)$,

$f_{P_2}(x) = \int_{U_2/U_2\cap\Gamma} f(xu)du$, define

$f_{P_2} \sim 0 \Longleftrightarrow \int_{M_2/\Gamma_2} (\psi(m), f_{P_2}(ma))dm = 0$ for all $a \in A_2$, all $\psi \in {}^oL^2(M_2/\Gamma_2,\sigma_2,\chi_2)$ (for all χ_2).

Consider now

$$J(a) = \int_{M_2/\Gamma_2} (\psi(m),\ E^o_\Lambda(ma))dm \qquad (a \in A_2),$$

where ψ is a function in some space ${}^oL^2(M_2/\Gamma_2,\sigma_2,\chi_2)$.

$$J(a) = \int_{{}^oP_2/\Gamma\cap P_2} (\psi(p), E_\Lambda(ap))dp = \int_{{}^oP_2/\Gamma\cap P_2} \sum_{\Gamma/\Gamma\cap P_1} (\psi(p),\varphi_\Lambda(ap\gamma))dp \ .$$

Fix $\gamma \in \Gamma$. Then $(\Gamma \cap P_2)\gamma(\Gamma \cap P_1)$ is the disjoint union of the cosets $\delta\gamma(\Gamma \cap P_1)$ where δ runs through $\Gamma \cap P_2$ modulo $\Gamma \cap P_2 \cap {}^\gamma P_1$. Hence

$$\int_{{}^oP_2/\Gamma\cap P_2} \sum_{(\Gamma\cap P_2)\gamma(\Gamma\cap P_1)/\Gamma\cap P_1} (\psi(p),\varphi_\Lambda(ap\delta))dp =$$

$$\int_{{}^oP_2/\Gamma\cap P_2} \sum_{\Gamma\cap P_2/\Gamma\cap P_2\cap{}^\gamma P_1} (\psi(p),\varphi_\Lambda(ap\delta\gamma))dp = \int_{{}^oP_2/\Gamma\cap P_2\cap{}^\gamma P_1} (\psi(p),\varphi_\Lambda(ap\gamma))dp \ .$$

Thus $J(a) = \sum_{\Gamma\cap P_2\backslash\Gamma/\Gamma\cap P_1} J_\gamma(a)$ with $J_\gamma(a) = \int_{{}^oP_2/\Gamma\cap P_2\cap{}^\gamma P_1} (\psi(p),\varphi_\Lambda(ap\gamma))dp.$

Since ψ is right-invariant under U_2 this leads us to consider

$$\Phi_{\Lambda,\gamma}(x) = \int_{U_2/U_2\cap\Gamma\cap{}^{\gamma}P_1} \varphi_\Lambda(xu\gamma)\,du$$

(to see that this integral exists one may write

$$\int_{U_2/U_2\cap\Gamma\cap{}^{\gamma}P_1} |\varphi_\Lambda(xu\gamma)|\,du = \int_{U_2/U_2\cap\Gamma} \sum_{\gamma^{-1}U_2\cap\Gamma/\gamma^{-1}U_2\cap\Gamma\cap P_1} |\varphi_\Lambda(xu\gamma\delta)|\,du < \infty).$$

Put $\Delta_\gamma = (U_2 \cap \Gamma \cap {}^{\gamma}P_1)(U_2 \cap {}^{\gamma}U_1)$; this is a group, because the first factor normalizes the second one, and it is closed, because the second factor is compact modulo its intersection with the first factor. Now, since the function $u \to \varphi_\Lambda(xu\gamma)$ on U_2 is right-invariant under $U_2 \cap {}^{\gamma}U_1$, the integral which defines $\Phi_{\Lambda,\gamma}$ can be written as an integral over U_2/Δ_γ times the volume of $\Delta_\gamma/U_2 \cap \Gamma \cap {}^{\gamma}P_1 = U_2 \cap {}^{\gamma}U_1/U_2 \cap \Gamma \cap {}^{\gamma}U_1$. The latter space being compact we give it volume 1 and obtain

$$\Phi_{\Lambda,\gamma}(x) = \int_{U_2/\Delta_\gamma} \varphi_\Lambda(xu\gamma)\,d\bar{u}\,.$$

We claim that $\Phi_{\Lambda,\gamma}(xv) = \Phi_{\Lambda,\gamma}(x)$ for $v \in {}^{o}P_2 \cap {}^{\gamma}U_1 = P_2 \cap {}^{\gamma}U_1$. Let $v \in P_2 \cap {}^{\gamma}U_1$. Then $\Phi_{\Lambda,\gamma}(xv) = \int_{U_2/\Delta_\gamma} \varphi_\Lambda(x\,{}^{v}u\,v\,\gamma)\,d\bar{u} = \int_{U_2/\Delta_\gamma} \varphi_\Lambda(x\,{}^{v}u\,\gamma)\,d\bar{u}$. Now v normalizes the groups U_2, Δ_γ (see below) and $U_2 \cap {}^{\gamma}U_1$ and conjugation by v does not change the Haar measures on U_2 and $U_2 \cap {}^{\gamma}U_1$, hence it does not change $d\bar{u}$, and we see that $\Phi_{\Lambda,\gamma}(xv) = \Phi_{\Lambda,\gamma}(x)$. We have still to check that ${}^{v}\Delta_\gamma = \Delta_\gamma$. It is enough to prove that ${}^{\delta}vv^{-1} \in \Delta_\gamma$ if $\delta \in U_2 \cap {}^{\gamma}P_1 \cap \Gamma$. To see this one observes that ${}^{\delta}vv^{-1} \in U_2$, so, since δ normalizes $P_2 \cap {}^{\gamma}U_1$, ${}^{\delta}vv^{-1} \in U_2 \cap (P_2 \cap {}^{\gamma}U_1) = U_2 \cap {}^{\gamma}U_1 \subset \Delta_\gamma$.

From the definitions of $J_\gamma(a)$ and $\Phi_{\Lambda,\gamma}(x)$ one gets immediately

$$J_\gamma(a) = \int_{M_2/\Delta_2} dm \int_{U_2/U_2\cap\Delta} (\psi(m),\varphi_\Lambda(amu\gamma))\,du = \int_{M_2/\Delta_2} (\psi(m),\Phi_{\Lambda,\gamma}(am))\,dm\,,$$

where $\Delta = P_2 \cap \Gamma \cap {}^{\gamma}P_1$, $\Delta_2 = \pi_{P_2|M_2}(\Delta)$. Put ${}^{*}\Gamma_2 = \pi_2({}^{o}\Gamma_2 \cap {}^{\gamma}P_1)$, ${}^{*}U_2 = \pi_2({}^{o}P_2 \cap {}^{\gamma}U_1)$ where $\pi_2 = \pi_{P_2|M_2}$. Δ_2 is a discrete subgroup of M_2 which normalizes ${}^{*}U_2$; by Lemma 30, ${}^{*}U_2/{}^{*}U_2 \cap \Delta_2$ is compact; so $\Delta_2{}^{*}U_2$ is a closed subgroup of M_2. From what we have seen above it follows that $\Phi_{\Lambda,\gamma}(x{}^{*}u) = \Phi_{\Lambda,\gamma}(x)$ for ${}^{*}u \in {}^{*}U_2$. So we get

$$J_\gamma(a) = \int_{M_2/\Delta_2{}^*U_2} d\bar{m} \int_{{}^*U_2/{}^*U_2\cap\Delta_2} (\psi(m^*u), \Phi_{\Lambda,\gamma}(am))d^*u .$$

Now $\Delta_2 \subset \Gamma_2$, so $\int_{{}^*U_2/{}^*U_2\cap\Delta_2} \psi(m^*u)d^*u = \int_{{}^*U_2/{}^*U_2\cap\Gamma_2} \psi(m^*u)d^*u$ up to a constant factor, and the latter integral is zero unless ${}^*U_2 = 1$, since ψ is a cuspidal form. Lemma 29 tells us that ${}^*U_2 = 1$ implies $A_2 \supset {}^\xi A_1$ for some $\xi \in G_{\mathbb{Q}}$. We summarize our results in two lemmas.

Lemma 31. Let $\varphi \in L^2(M_1/\Gamma_1,\sigma_1,\chi_1)$, $\Lambda \in (\mathfrak{a}^*_{1,c})^-$. Define E_Λ and E^0_Λ as above. Then $E^0_\Lambda \sim 0$ if rank $P_2 \leq$ rank P_1 and P_1, P_2 are not associated.

Lemma 32. Fix $\gamma \in \Gamma$ and $\psi \in {}^0L^2(M_2/\Gamma_2,\sigma_2,\chi_2)$ and define $J_\gamma(a)$ as above. Then $J_\gamma = 0$ unless $P_2 \cap {}^\gamma U_1 \subset U_2$.

Under the assumption that φ is a cusp form we can prove more.

Lemma 33. Let $\varphi \in {}^0L^2(M_1/\Gamma_1,\sigma_1,\chi_1)$ and $\Lambda \in (\mathfrak{a}^*_{1,c})^-$. Define $\Phi_{\Lambda,\gamma}$ as above. Then $\Phi_{\Lambda,\gamma} = 0$ unless ${}^{\gamma^{-1}}U_2 \cap P_1 \subset U_1$.

Proof. $\Phi_{\Lambda,\gamma}(x) = \int_{U_2/U_2\cap{}^\gamma P_1} d\bar{u} \int_{U_2\cap{}^\gamma P_1/U_2\cap{}^\gamma P_1\cap\Gamma} \varphi_\Lambda(xuu_1\gamma)du_1$.

$$\int_{U_2\cap{}^\gamma P_1/U_2\cap{}^\gamma P_1\cap\Gamma} \varphi_\Lambda(xu_1\gamma)du_1 = \int_{{}^{\gamma^{-1}}U_2\cap P_1/{}^{\gamma^{-1}}U_2\cap P_1\cap\Gamma} \varphi_\Lambda(x\gamma u_2)du_2 =$$

$$\int_{{}^*U_1/\pi_1({}^{\gamma^{-1}}U_2\cap P_1\cap\Gamma)} \varphi_\Lambda(x\gamma^*u)d^*u, \text{ where } {}^*U_1 = \pi_1({}^{\gamma^{-1}}U_2 \cap P_1),\ \pi_1 = \pi_{P_1|M_1} .$$

Since $\pi_1({}^{\gamma^{-1}}U_2 \cap P_1 \cap \Gamma) \subset \Gamma_1$ and $\varphi_\Lambda(x\gamma^*u) = \varphi(x\gamma^*u)e^{(\Lambda-\varrho_1)(H_1(x\gamma))}$, the latter integral is zero unless ${}^*U_1 = 1$, i.e. ${}^{\gamma^{-1}}U_2 \cap P_1 \subset U_1$.

Corollary 1. If $\varphi \in {}^0L^2(M_1/\Gamma_1,\sigma_1,\chi_1)$, then $E^0_\Lambda = 0$ unless ${}^{\gamma^{-1}}U_2 \cap P_1 \subset U_1$ for some $\gamma \in \Gamma$.

Proof. $E^0_\Lambda(x) = \int_{U_2/U_2\cap\Gamma} \sum_{\Gamma/\Gamma\cap P_1} \varphi_\Lambda(xu\gamma)du$; and for any $\gamma \in \Gamma$:

$$\int_{U_2/U_2\cap\Gamma} \sum_{(\Gamma\cap U_2)\gamma(\Gamma\cap P_1)/\Gamma\cap P_1} \varphi_\Lambda(xu\delta)\,du = \int_{U_2/U_2\cap\Gamma} \sum_{U_2\cap\Gamma/U_2\cap\Gamma\cap{}^\gamma P_1} \varphi_\Lambda(xu\delta\gamma)\,du =$$

$$\int_{U_2/U_2\cap\Gamma\cap{}^\gamma P_1} \varphi_\Lambda(xu\gamma)\,du = \Phi_{\Lambda,\gamma}(x).$$

Corollary 2. $E_\Lambda^o = o$ if rank $P_2 \geq$ rank P_1 and P_1, P_2 are not associated.

Proof. This follows from Corollary 1, since ${}^{\gamma-1}U_2 \cap P_1 \subset U_1$ implies the existence of an element ξ of $G_{\mathbb{Q}}$ such that $A_1 \supset {}^\xi A_2$ (Lemma 29).

Corollary 3. Let $\varphi \in {}^oL^2(M_1/\Gamma_1,\sigma_1,\chi_1)$ and $\psi \in {}^oL^2(M_2/\Gamma_2,\sigma_2,\chi_2)$. Define $J_\gamma(a)$ as above. Then $J_\gamma = o$ unless $P_2 \cap {}^\gamma U_1 \subset U_2$, $P_1 \cap {}^{\gamma-1}U_2 \subset U_1$. Hence, if $J_\gamma \neq o$ for some $\gamma \in \Gamma$, then P_1 and P_2 are associated.

Proof. This follows immediately from Lemmas 32 and 33.

§5. The constant term of E_Λ (P_1 and P_2 associate).

We use the notations of §4 and assume now that P_1 and P_2 are associate.

For any $s \in w(\mathfrak{a}_1,\mathfrak{a}_2)$ choose $y \in G_{\mathbb{Q}}$ such that $Ad(y) = s$ on $\mathfrak{a}_1$. Then P_2yP_1 depends on s and not on y: define $\Gamma(s) = \Gamma \cap P_2yP_1$. The sets $\Gamma(s)$, $s \in w(\mathfrak{a}_1,\mathfrak{a}_2)$ are disjoint, and if P_1 and P_2 are mincuspidal, the union of the $\Gamma(s)$ is Γ (this can easily be derived from the fact that $N(A_1) \cap U_1^-U_1 = 1$, and Bruhat's lemma).

Lemma 34. Let $\varphi \in {}^oL^2(M_1/\Gamma_1,\sigma_1,\chi_1)$ and $\gamma \in \Gamma(s)$. Then

$$\Phi_{\Lambda,\gamma}(ma) = e^{(s\Lambda-\varrho_2)(\log a)}\,\Phi_{\Lambda,\gamma}(m) \qquad (m \in M_2,\ a \in A_2).$$

Proof. Choose a mincuspidal pair $(P,A) \prec (P_1,A_1)$ and fix $\xi \in G_{\mathbb{Q}}$ such that $(P_2,A_2) \succ {}^\xi(P,A)$. Then $(P_1,A_1) = (P_{F_1},A_{F_1})$ and $(P_2,A_2) = {}^\xi(P_{F_2},A_{F_2})$ with $F_i \subset \Sigma^o(P|A)$. We may assume that $\Phi_{\Lambda,\gamma} \neq o$. Then $U_2 \cap {}^\gamma P_1 \subset {}^\gamma U_1$ (Lemma 33). Write $\xi^{-1}\gamma = u_o w p_o$ with $u_o \in U_{\mathbb{Q}}$, $w \in N(A)_{\mathbb{Q}}$, $p_o \in P_{\mathbb{Q}}$. Then $U_{F_2} \cap {}^wP_{F_1} \subset {}^wU_{F_1}$, hence ${}^wA_{F_1} = A_{F_2}$ (cf. the end of the proof of Lemma 29). We saw in §4 that

$\Phi_{\Lambda,\gamma}(x) = \int_{U_2/\Delta_\gamma} \varphi_\Lambda(xu\gamma)d\bar{u}$; in the present case $\Delta_\gamma = U_2 \cap {}^\gamma U_1$, and so

$$\Phi_{\Lambda,\gamma}(x) = \int_{U_{F_2}/U_{F_2}\cap {}^w U_{F_1}} \varphi_\Lambda(x\, {}^{\xi u_o}u\gamma)d\bar{u} .$$

We have $ma\, {}^{\xi u_o}u\gamma = m\xi a_2 u_o uwp_o$, where $a_2 = {}^{\xi^{-1}}a \in A_{F_2}$ $(m \in M_2,\ a \in A_2)$. Writing $u_o = u_1 u_2$ with $u_1 \in M_{F_2} \cap U_{\mathbb{Q}}$, $u_2 \in (U_{F_2})_{\mathbb{Q}}$ we get

$$\Phi_{\Lambda,\gamma}(ma) = \int \varphi_\Lambda(m\xi a_2 u_1 uwp_o)d\bar{u} .$$

Now we have $m\xi a_2 u_1 uwp_o = m\xi u_1\, {}^{a_2}uw^{w^{-1}}a_2 p_o \in m\xi u_1\, {}^{a_2}uwp_o\, {}^{w^{-1}}a_2\, U_{F_1}$, which gives

$$\Phi_{\Lambda,\gamma}(ma) = \beta(a_2) \int_{U_{F_2}/U_{F_2}\cap {}^w U_{F_1}} \varphi_\Lambda(m\xi u_1\, {}^{a_2}uwp_o)d\bar{u} ,$$

where $\beta(a_2) = e^{(\Lambda-\varrho_1)(\log\, {}^{w^{-1}}a_2)} = e^{(s_o(\Lambda-\varrho_1))(\log a_2)}$, s_o being the element of $w(\mathfrak{a}_{F_1},\mathfrak{a}_{F_2})$ defined by w . Since $a_2 \in A_{F_2} = {}^w A_{F_1}$, it normalizes both U_{F_2} and $U_{F_2} \cap {}^w U_{F_1}$. Put

$$\delta(a_2) = \det_{\mathfrak{n}_{F_2}/\mathfrak{n}_{F_2}\cap {}^w\mathfrak{n}_{F_1}}(\mathrm{Ad}(a_2)).$$

Then

$$\Phi_{\Lambda,\gamma}(ma) = \beta(a_2)\delta(a_2)^{-1} \int_{U_{F_2}/U_{F_2}\cap {}^w U_{F_1}} \varphi_\Lambda(m\xi u_1 uwp_o)d\bar{u}$$

and this will prove the lemma if we show that $\beta(a_2)\delta(a_2)^{-1}$ is equal to the factor in Lemma 34.

If $\alpha \in \Sigma_1 = \Sigma_{F_1} = \Sigma(P_{F_1}|A_{F_1})$, then either $s_o\alpha$ or $-s_o\alpha$ belongs to $\Sigma_{F_2} = \Sigma(P_{F_2}|A_{F_2})$. Let Σ_1' , resp. Σ_1'' , be the set of roots α in Σ_1 such that $s_o\alpha \in \Sigma_{F_2}$, resp. $-s_o\alpha \in \Sigma_{F_2}$. Then $\Sigma_1 = \Sigma_1' \cup \Sigma_1''$, $\Sigma_{F_2} = s_o\Sigma_1' \cup -s_o\Sigma_1''$. Put

$$\mathfrak{n}_1' = \sum_{\alpha\in\Sigma_1'} \mathfrak{n}_{1,\alpha} \ , \mathfrak{n}_1'' = \sum_{\alpha\in\Sigma_1''} \mathfrak{n}_{1,\alpha} .$$

Then $\mathfrak{n}_1 = \mathfrak{n}_1' + \mathfrak{n}_1''$, $\mathfrak{n}_{F_2} = {}^w\mathfrak{n}_1' + {}^w(\mathfrak{n}_1'')^-$, $\mathfrak{n}_{F_2} \cap {}^w\mathfrak{n}_1 = {}^w\mathfrak{n}_1'$ and , if $\varrho_1' = \frac{1}{2}\,\mathrm{tr}_{\mathfrak{n}_1'}(\mathrm{ad}\, H)$, $\varrho_1'' = \frac{1}{2}\,\mathrm{tr}_{\mathfrak{n}_1''}(\mathrm{ad}\, H)$ $(H \in \mathfrak{a}_1)$, then $\varrho_1 = \varrho_1' + \varrho_1''$.

$\varrho_{F_2} = s_o\varrho_1' - s_o\varrho_1''$. Now $\mathrm{tr}_{\mathfrak{n}_{F_2}/^w\mathfrak{n}_1'}(\mathrm{ad}\ H) = 2(\varrho_{F_2} - s_o\varrho_1') = -2s_o\varrho_1''$ for $H \in \mathfrak{a}_{F_2}$. Hence $\beta(a_2)\delta(a_2)^{-1} = \exp\{(s_o(\Lambda-\varrho_1) + 2s_o\varrho_1'')(\log a_2)\} = \exp\{(s_o\Lambda-\varrho_{F_2})(\log a_2)\}$. Call s' the element of $w(\mathfrak{a}_1,\mathfrak{a}_2)$ defined by ξw. From $\gamma = {}^{\xi}u_o\ \xi w\ p_o$ one sees that $\gamma \in \Gamma(s')$, so $s' = s$ and $\beta(a_2)\delta(a_2)^{-1} = \exp\{(s\Lambda-\varrho_2)(\log a)$, which completes the proof of Lemma 34.

Fix $\varphi \in {}^oL^2(M_1/\Gamma_1,\sigma_1,\chi_1)$. Then $E_\Lambda^o(x) = \sum_{\Gamma\cap U_2\backslash\Gamma/\Gamma\cap P_1} \Phi_{\Lambda,\gamma}(x)$ (see the proof of Corollary 1 of Lemma 33). During the proof of Lemma 34 we saw that $\Phi_{\Lambda,\gamma} \neq 0$ implies $\gamma \in \Gamma(s)$ for some $s \in w(\mathfrak{a}_1,\mathfrak{a}_2)$, so

$$E_\Lambda^o(x) = \sum_{s\in w(\mathfrak{a}_1,\mathfrak{a}_2)} \Phi_{\Lambda,s}(x)$$

with

$$\Phi_{\Lambda,s}(x) = \sum_{\Gamma\cap U_2\backslash\Gamma(s)/\Gamma\cap P_1} \Phi_{\Lambda,\gamma}(x) = \int_{U_2/U_2\cap\Gamma} \sum_{\Gamma(s)/\Gamma\cap P_1} \varphi_\Lambda(xu\gamma)\,du\ .$$

<u>Lemma 35</u>. Fix $\gamma_o \in \Gamma$. Put

$$\Psi_{\Lambda,\gamma_o}(x) = \int_{U_2/U_2\cap\Gamma} \sum_{(\Gamma\cap P_2)\gamma_o(\Gamma\cap\Gamma_1)/\Gamma\cap\Gamma_1} \varphi_\Lambda(xu\gamma)\,du\ .$$

Then the function $m \to \Psi_{\Lambda,\gamma_o}(m)$ on M_2 lies in ${}^o\mathcal{A}(M_2/\Gamma_2,\sigma_2)$.

<u>Proof</u>. It is obvious that the function lies in $\mathcal{A}(M_2/\Gamma_2,\sigma_2)$. Let $({}^*P,{}^*A)$ be a cuspidal pair of M_2 and ${}^*P \neq M_2$. Let P' be the corresponding cuspidal subgroup of G such that $P' \subset P_2$ (Lemma 2). By means of the integration formula

$$\int_{U'/U'\cap\Gamma} f(u')\,du' = \int_{{}^*U/{}^*U\cap\Gamma_2} d{}^*u \int_{U_2/U_2\cap\Gamma} f({}^*uu)\,du \qquad (f \in C(U'/U'\cap\Gamma))$$

we see that

$$\int_{U'/U'\cap\Gamma} \sum_{(\Gamma\cap P_2)\gamma_o(\Gamma\cap P_1)/\Gamma\cap P_1} \varphi_\Lambda(xu'\gamma)\,du' = \int_{{}^*U/{}^*U\cap\Gamma_2} \Psi_{\Lambda,\gamma_o}(x{}^*u)\,d{}^*u\ .$$

Here the left hand side is a sum of terms of the form

$$\int_{U'/U'\cap\Gamma} \sum_{(\Gamma\cap U')\delta(\Gamma\cap P_1)/\Gamma\cap P_1} \varphi_\Lambda(xu'\gamma)\,du' = \int_{U'/U'\cap\Gamma\cap{}^{\delta}P_1} \varphi_\Lambda(xu'\delta)\,du' ,$$

and this is zero by Lemma 33, since rank $P' >$ rank P_1 .

From Lemma 35 we see that $\Phi_{\Lambda,s}|_{M_2} \in {}^{o}\mathcal{A}(M_2/\Gamma_2,\sigma_2)$. We want to find out how $\mathfrak{Z}_{M_2}$ acts on that function.

Denoting by $\mathcal{H}$ the set of all characters of $\mathfrak{Z}$ we put the following definition:

$${}^{o}L^2(G/\Gamma,\sigma,\xi) = \sum_{\chi\in\xi} {}^{o}L^2(G/\Gamma,\sigma,\chi)$$

when ξ is a finite subset of $\mathcal{H}$. For any representation χ' (commuting with σ) of $\mathfrak{Z}$ on V there exists a finite subset ξ of $\mathcal{H}$ such that ${}^{o}L^2(G/\Gamma,\sigma,\chi') =$ $= {}^{o}L^2(G/\Gamma,\sigma,\xi)$. In particular this can be applied to the group M , if $P = MAU$ is a cuspidal group. It will then be permitted to consider the spaces ${}^{o}L^2(M/\Gamma_M,\sigma_M,\xi)$ ($\xi \subset \mathcal{H}_M$ = set of all characters of the center $\mathfrak{Z}_M$ of the enveloping algebra of $\mathfrak{m}_c$, ξ finite) instead of ${}^{o}L^2(M/\Gamma_M,\sigma_M,\chi)$ (χ a representation of $\mathfrak{Z}_M$) .

Returning to our associate cuspidal groups P_1, P_2 we write $\mathfrak{Z}_{M_i} = \mathfrak{Z}_i$, $\mathcal{H}_{M_i} = \mathcal{H}_i$. Let s be an element of $w(\mathfrak{a}_1,\mathfrak{a}_2)$. Choose $y \in G_{\mathbb{Q}}$ such that $\mathrm{Ad}(y) = s$ on $\mathfrak{a}_1$. Then ${}^{y}\mathfrak{m}_1 = \mathfrak{m}_2$, so ${}^{y}\mathfrak{Z}_1 = \mathfrak{Z}_2$, and since any other element of $G_{\mathbb{Q}}$ which induces s is of the form yma with $m \in M_1$, $a \in A_1$, the isomorphism $\mathfrak{Z}_1 \to \mathfrak{Z}_2$ defined by y does not depend on y but on s only (because of the hypothesis 2')). So we may write ${}^{s}\zeta$ for the image of ζ in $\mathfrak{Z}_2$ if $\zeta \in \mathfrak{Z}_1$. Finally s defines an isomorphism $\mathcal{H}_1 \to \mathcal{H}_2$ (${}^{s}\chi({}^{s}\zeta) = \chi(\zeta)$ for $\chi \in \mathcal{H}_1$, $\zeta \in \mathfrak{Z}_1$).

Lemma 36. Let $\varphi \in {}^{o}L^2(M_1/\Gamma_1,\sigma_1,\chi)$, $\chi \in \mathcal{H}_1$. For any $\Lambda \in (\mathfrak{a}^*_{1,c})^-$ and $s \in w(\mathfrak{a}_1,\mathfrak{a}_2)$ the function $m \to \Phi_{\Lambda,s}(m)$ on M_2 lies in ${}^{o}L^2(M_2/\Gamma_2,\sigma_2,{}^{s}\chi)$.

Proof. Extend $\mathfrak{a}_1$ to a Cartan subalgebra $\mathfrak{h}_1$ of $\mathfrak{g}$. Then $\mathfrak{h}_1 = \mathfrak{a}_1 + \mathfrak{b}_1$, where $\mathfrak{b}_1 = \mathfrak{m}_1 \cap \mathfrak{h}_1$; $\mathfrak{b}_1$ is a Cartan subalgebra of $\mathfrak{m}_1$. Let γ_1 be the canonical isomorphism of $\mathfrak{Z}_1$ on the algebra $I(\mathfrak{b}_{1,c})$ of invariants of the Weyl group $W_1^o = W(\mathfrak{m}_1|\mathfrak{b}_1)$ in the symmetrical algebra $S(\mathfrak{b}_{1,c})$. By transport of structure χ defines a character of $I(\mathfrak{b}_{1,c})$. Hence, from the structure of $I(\mathfrak{b}_{1,c})$, there is an element λ in $\mathfrak{b}^*_{1,c}$ such that $\chi(\zeta) = \gamma_1(\zeta)(\lambda)$ for $\zeta \in \mathfrak{Z}_1$.

The group W_1^o is a subgroup of the Weyl group W_1 of $(\mathfrak{g}_1, \mathfrak{f}_1)$ and for $\tau \in W_1$ we have $\tau \in W_1^o$ if and only if τ leaves $\mathfrak{a}_1$ pointwise fixed. In the sequel we extend linear functions on $\mathfrak{a}_{1,c}$ (resp. $\mathfrak{b}_{1,c}$) to $\mathfrak{f}_{1,c}$ by defining them zero on $\mathfrak{b}_{1,c}$ (resp. $\mathfrak{a}_{1,c}$) and for $\Lambda \in \mathfrak{a}^*_{1,c}$ we put $\Lambda_1 = \Lambda + \lambda \in \mathfrak{f}^*_{1,c}$.

We saw in §2 that, for $z \in \mathfrak{Z}$, $z\varphi_\Lambda = \varphi_\Lambda \chi(z:\Lambda)$ where $\chi(z:\Lambda) = \chi(\mu_1(z:\Lambda))$, μ_1 is the canonical homomorphism $\mathfrak{Z} \to \mathfrak{Z}_1\mathfrak{A}_1$ and $\mu_1(z:\Lambda)$ is the image of $\mu_1(z)$ under the homomorphism $\mathfrak{Z}_1\mathfrak{A}_1 = \mathfrak{Z}_1 \otimes_C \mathfrak{A}_1 \to \mathfrak{Z}_1$ induced by Λ.

Now we have the following easy lemma which we shall not prove here.

<u>Lemma 37</u>. Let Φ be a C^∞-function on G which satisfies the following conditions.

1) $\Phi(kamu) = e^{(s\Lambda-\rho_2)(\log a)}\sigma(k)\Phi(m)$ $\quad (k \in K,\ a \in A_2,\ m \in M_2,\ u \in U_2)$;

2) $z\Phi = \chi(z:\Lambda)\Phi$ $\quad (z \in \mathfrak{Z})$.

Assume that for every $\tau \in W_1 - W_1^o$, $\tau\Lambda_1 - \Lambda$ is not identically zero on $\mathfrak{a}_1$. Then $\Phi(m;\zeta) = {}^s\chi(\zeta)\Phi(m)$ $\quad (\zeta \in \mathfrak{Z}_2,\ m \in M_2)$.

Take in this lemma $\Phi = \Phi_{\Lambda,s}$. From the definitions and Lemma 34 one sees that condition 1) is satisfied and from the results of §2 that condition 2 is satisfied. When is the assumption on Λ true ?

Suppose that $\tau\Lambda_1 - \Lambda = 0$ on $\mathfrak{a}_1$ for <u>all</u> $\Lambda \in \mathfrak{a}^*_{1,c}$. Then $\tau\Lambda - \Lambda = 0$ on $\mathfrak{a}_1$ for all Λ, hence $\tau^{-1}H - H \in \mathfrak{b}_1$ for all $H \in \mathfrak{a}_1$, but then $\tau^{-1}H = H$, since $\mathfrak{a}_1$ is orthogonal to $\mathfrak{b}_1$ and $\tau^{-1}H$ and H have the same length. So $\tau \in W_1^o$. This means that, if $\tau \notin W_1^o$, the set V_τ of all $\Lambda \in \mathfrak{a}^*_{1,c}$ such that $\tau\Lambda_1 - \Lambda = 0$ on $\mathfrak{a}_1$ is a proper linear subvariety of $\mathfrak{a}^*_{1,c}$. The assumption of Lemma 37 is true if $\Lambda \notin V_\tau$ for each $\tau \in W_1 - W_1^o$. This proves that $\Phi_{\Lambda,s}(m;\zeta) = {}^s\chi(\zeta)\Phi_{\Lambda,s}(m)$ if $\Lambda \in (\mathfrak{a}^*_{1,c})^- \cap \bigcap {}^c V_\tau$. By continuity the same is true for all $\Lambda \in (\mathfrak{a}^*_{1,c})^-$.

So $\Phi_{\Lambda,s}\big|_{M_2} \in {}^o\mathcal{A}(M_2/\Gamma_2, \sigma_2, {}^s\chi) = {}^oL^2(M_2/\Gamma_2, \sigma_2, {}^s\chi)$.

Thus we have obtained a linear map ${}^oL^2(M_1/\Gamma_1, \sigma_1, \chi) \to {}^oL^2(M_2/\Gamma_2, \sigma_2, {}^s\chi)$, viz. $\varphi \mapsto \Phi_{\Lambda,s}\big|_{M_2}$, and so we have linear maps ${}^oL^2(M_1/\Gamma_1, \sigma_1, \xi) \to {}^oL^2(M_2/\Gamma_2, \sigma_2, {}^s\xi)$ for

$\xi \subset \mathcal{H}_1$, ξ finite. We state the results we have obtained in a theorem.

<u>Theorem 5</u>. Let ξ be a finite subset of $\mathcal{H}_1$. Fix $s \in w(\mathfrak{a}_1, \mathfrak{a}_2)$, $\Lambda \in (\mathfrak{a}^*_{1,c})^-$. There exists a unique linear transformation $c(s:\Lambda)$ from $L(\xi) = {}^0L^2(M_1/\Gamma_1, \sigma_1, \xi)$ to $L({}^s\xi) = {}^0L^2(M_2/\Gamma_2, \sigma_2, {}^s\xi)$ such that

$$\int_{U_2/U_2\cap\Gamma} \sum_{\Gamma(s)/\Gamma\cap P_1} \varphi_\Lambda(mau\gamma)\,du = e^{(s\Lambda-\varrho_2)(\log a)} (c(s:\Lambda)\varphi)(m)$$

for $\varphi \in L(\xi)$, $m \in M_2$, $a \in A_2$. Moreover $c(s:\Lambda)$ is holomorphic in Λ for $\Lambda \in (\mathfrak{a}^*_{1,c})^-$. Finally

$$\int_{U_2/U_2\cap\Gamma} \sum_{\Gamma/\Gamma\cap P_1} \varphi_\Lambda(mau\gamma)\,du = \sum_{s\in w(\mathfrak{a}_1,\mathfrak{a}_2)} e^{(s\Lambda-\varrho_2)(\log a)} (c(s:\Lambda)\varphi)(m) .$$

We denote by $|c(s:\Lambda)|$ the norm of the linear transformation $c(s:\Lambda)$.

<u>Lemma 38</u>. There exist a number c and an element H_1 of $\mathfrak{a}_1$ such that

$$|c(s:\Lambda)| \leq c \frac{e^{(\Lambda_R-\varrho_1)(H_1)}}{\prod_{\alpha\in\Sigma^0} |\langle\Lambda_R+\varrho_1,\alpha\rangle|}$$

for $\Lambda \in (\mathfrak{a}^*_{1,c})^-$, $\Lambda_R = \mathrm{Re}(\Lambda)$, $s \in w(\mathfrak{a}_1, \mathfrak{a}_2)$.

<u>Proof</u>. Put $T_m(\psi) = \psi(m)$ for $\psi \in L({}^s\xi)$, $m \in M_2$. Choose $m_1,\dots,m_r$ in M_2 such that $T_{m_1},\dots,T_{m_r}$ generate $\mathrm{Hom}(L({}^s\xi),V)$ as a module over $\mathrm{End}(V)$; this is possible because $T_m(\psi) = 0$ for all m implies $\psi = 0$. Choose a base $\varphi_1,\dots,\varphi_l$ for $L(\xi)$. There is a constant c such that for all linear mappings C of $L(\xi)$ into $L({}^s\xi)$ we have $|C| \leq c \sup_{i,j} |T_{m_j}(C\varphi_i)|$.

Fix a compact set ω in U_2 such that $\omega(\Gamma \cap U_2) = U_2$. Choose a compact set Ω in G such that $m_j\omega \subset \Omega$ $(1 \leq j \leq r)$. If φ is one of the φ_i and m is one of the m_j , then

$$|(c(s:\Lambda)\varphi)(m)| \leq \int_{U_2/U_2\cap\Gamma} \sum_{\Gamma/\Gamma\cap P_1} |\varphi_\Lambda(mu\gamma)|\,du \leq$$

$$\sup_{u\in\omega} \sum_{\Gamma/\Gamma\cap P_1} |\varphi_\Lambda(mu\gamma)| \leq \sup_{x\in\Omega} \sum_{\Gamma/\Gamma\cap P_1} |\varphi_\Lambda(x\gamma)| \leq$$

$$\|\varphi\|_\infty \sup_x \sum_{\Gamma/\Gamma\cap P_1} e^{(\Lambda_R-\varrho_1)(H_1(x\gamma))} \leq c\,\frac{e^{(\Lambda_R-\varrho_1)(H_1)}}{\Pi|\langle\Lambda_R+\varrho_1,\alpha\rangle|}\;;$$

the last inequality follows from Lemma 23 as it has been shown in Remark 2) at the end of §2.

§6. Orthogonality and density of the E_f .

If $P = MAU$ is a cuspidal subgroup of G , σ a representation of K and ξ a finite subset of $\mathcal{H}_M$, we define

$$^{\circ}\mathcal{D}_{P|A}(\sigma,\xi) = \sum_{\chi\in\xi} {}^{\circ}\mathcal{D}_{P|A}(\sigma,\chi)\,.$$

The linear space $^{\circ}\mathcal{D}_{P|A}(\sigma,\xi)$ is spanned by the functions f on $G/U(\Gamma\cap P)$ of the form $f(x) = \varphi(x)v(a(x))$ with $\varphi \in {}^{\circ}L^2(M/\Gamma_M,\sigma_M,\xi)$, $v \in C_c^\infty(A)$ (see §3).

Let P_1, P_2 be cuspidal subgroups with decompositions $P_i = M_iA_iU_i$. Let ξ_i be a finite subset of $\mathcal{H}_{M_i}$ $(i = 1,2)$ and write $^{\circ}\mathcal{D}_i(\sigma,\xi_i) = {}^{\circ}\mathcal{D}_{P_i|A_i}(\sigma,\xi_i)$.

<u>Lemma 39</u>. If $f_i \in {}^{\circ}\mathcal{D}_i(\sigma,\xi_i)$ $(i = 1,2)$, then $((E_{f_2},E_{f_1}))_{G/\Gamma} = 0$ unless P_1, P_2 are associate.

<u>Proof</u>. By Lemma 27 one has

$$((E_{f_2},E_{f_1}))_{G/\Gamma} = \int_{A_2} e^{2\varrho_2(\log a)}\,da \int_{M_2/\Gamma_2} (f_2(am),\, E^{\circ}_{f_1}(am))\,dm,$$

where $\Gamma_2 = \Gamma_{M_2}$, and from Lemma 28 it follows that

$$E^{\circ}_{f_1}(x) = \int_{\mathfrak{a}_1^*} E^{\circ}(\Lambda:\hat{f}_1(\Lambda):x)\,d\Lambda_I \qquad (\Lambda = \Lambda_R + i\Lambda_I,\ \Lambda_R \in (\mathfrak{a}_1^*)^-)\ .$$

By Corollary 2 of Lemma 33, $E^{o}(\Lambda:\hat{f}_1(\Lambda)) = o$ if rank $P_1 \leq$ rank P_2 and P_1, P_2 are not associate. This proves Lemma 39 in the case that rank $P_1 \leq$ rank P_2. Hence it is true in general.

Lemma 40. Assume P_1 and P_2 are associate. Then for $f_i \in {}^{o}\mathcal{D}_i(\sigma,\xi_i)$ $(i = 1,2)$ we have

$$((E_{f_2},E_{f_1}))_{G/\Gamma} = \sum_{s\in w(\mathfrak{a}_1,\mathfrak{a}_2)} \int_{\mathfrak{a}_1^*} ((\hat{f}_2(-s\bar{\Lambda}),\ c(s:\Lambda)\hat{f}_1(\Lambda)))_{M_2/\Gamma_2}\, d\Lambda_I$$

where $\Lambda_R \in (\mathfrak{a}_1^*)^-$.

Proof. We use the same formulas as in the proof of Lemma 39. Theorem 5 gives

$$E^{o}(\Lambda:\hat{f}_1(\Lambda):ma) = \sum_{s\in w(\mathfrak{a}_1,\mathfrak{a}_2)} e^{(s\Lambda-\varrho_2)(\log a)}(c(s:\Lambda)\varphi_1)(m)$$

where $\varphi_1 = \hat{f}_1(\Lambda) \in {}^{o}L^2(M_1/\Gamma_1,\sigma_1,\xi_1)$. Suppose that $f_2(x) = \varphi_2(x)v(a_2(x))$, $\varphi_2 \in {}^{o}L^2(M_2/\Gamma_2,\sigma_2,\xi_2)$, $v \in C_c^{\infty}(A_2)$. Then

$$(f_2(am),E_{f_1}^{o}(am)) = \sum_s \int_{\mathfrak{a}_1^*} e^{(s\Lambda-\varrho_2)(\log a)}\ \overline{v(a)}\ (\varphi_2(m),(c(s:\Lambda)\varphi_1)(m))\ d\Lambda_I ,$$

$$((E_{f_2},E_{f_1}))_{G/\Gamma} = \sum_s \int_{\mathfrak{a}_1^*} \overline{\hat{v}(-s\bar{\Lambda})}\ ((\varphi_2,\ c(s:\Lambda)\varphi_1))_{M_2/\Gamma_2}\ d\Lambda_I ,$$

which gives the wanted formula.

N.B. We have interchanged integration over $\mathfrak{a}_1^*$ and $A_2 \times M_2/\Gamma_2$. The integral over $\mathfrak{a}_1^* \times A_2 \times M_2/\Gamma_2$ is absolutely convergent as it obviously follows from the observation that

$$\int_{\mathfrak{a}_1^*} \|c(s:\Lambda)\hat{f}_1(\Lambda)\|_{M_2/\Gamma_2}\ d\Lambda_I < \infty \qquad (\Lambda_R \in (\mathfrak{a}_1^*)^-).$$

For any cuspidal pair (P,A) we denote by $E_{P|A}(\sigma,\chi)$, resp. $E_{P|A}(\sigma,\xi)$, the space of all functions E_f with $f \in {}^{o}\mathcal{D}_{P|A}(\sigma,\chi)$, resp. ${}^{o}\mathcal{D}_{P|A}(\sigma,\xi)$, if $\chi \in \mathcal{H}_M$, $\xi \subset \mathcal{H}_M$, ξ finite. Then $E_{P|A}(\sigma,\xi)$ is a subspace of $L^2(G/\Gamma,\sigma)$ (Cor. of Lemma 26).

Lemma 41. Let α be a function in $C_c(G)$ such that ${}^{k}\alpha = \alpha$ for $k \in K$ (i.e. $\alpha(k^{-1}xk) = \alpha(x)$). Then $\alpha * f \in {}^{o}\mathcal{D}_{P|A}(\sigma,\chi)$ for $f \in {}^{o}\mathcal{D}_{P|A}(\sigma,\chi)$ and $\alpha * E_f = E_{\alpha * f}$.

<u>Proof</u>. Put $\beta(x) = \int\int_{K\times U} \alpha((kxu)^{-1})\ \sigma(k)\ dk\ du$. Then, if $f \in {}^{o}\mathcal{D}_{P|A}(\sigma,\chi)$ and $f' = \alpha * f$,

$$f'(p) = \int\int_{M\times A} \beta(ma)\ f(map)\ e^{2\varrho(\log a)}\ dm\ da \quad (p \in P).$$

Of course $f'(kxu) = \sigma(k)f'(x)$ for $k \in K$, $u \in U$. Assuming that $f(x) = \varphi(x)v(a(x))$ with $\varphi \in {}^{o}L^2(M/\Gamma_M,\sigma_M,\chi)$, $v \in C_c^{\infty}(A)$, we have

$$f'(m_o a_o) = \iint \beta(ma)\varphi(mm_o)v(aa_o)e^{2\varrho(\log a)}dm\ da \quad (m_o \in M,\ a_o \in A).$$

Clearly the function $m_o \to \int_M \beta(ma)\varphi(mm_o)dm$ lies in ${}^{o}L^2(M/\Gamma_M,\sigma_M,\chi)$, so, if $\varphi_1,\dots,\varphi_N$ is a base for that space, then

$$\int_M \beta(ma)\varphi(mm_o)dm = \sum_{1\leq i\leq N} v_i(a)\ \varphi_i(m_o) ,$$

where the v_i are functions belonging to $C_c(A)$. Hence

$$f'(m_o a_o) = \sum \varphi_i(m_o)(v_i' * v)(a_o) ,$$

where we have put $v_i'(a) = v_i(a^{-1})e^{-2\varrho(\log a)}$, $v_i' * v \in C_c^{\infty}(A)$. This proves that $f' \in {}^{o}\mathcal{D}_{P|A}(\sigma,\chi)$. The second assertion of the lemma is then trivial.

<u>Lemma 42</u>. Let $\psi \in L^2(G/\Gamma,\sigma)$. Assume ψ is orthogonal to $E_{P|A}(\sigma,\chi)$ for all cuspidal pairs (P,A) (including $(G,1)$) and all $\chi \in \mathcal{H}_M$. Then $\psi = o$.

<u>Proof</u>. Choose a function α in $C_c(G)$ such that ${}^{k}\alpha = \alpha$ for $k \in K$. The function $\psi_1 = \alpha * \psi$ has the same properties as ψ , because

$$((\psi_1,E_f)) = ((\alpha*\psi,E_f)) = ((\psi,\tilde{\alpha}*E_f)) = ((\psi,E_{\tilde{\alpha}*f}))$$

(here $\tilde{\alpha}(x) = \overline{\alpha(x^{-1})}$). Moreover $\psi_1 \in \mathcal{A}(G/\Gamma,\sigma)$, so, by Lemma 26,

$$((\psi_1,E_f))_{G/\Gamma} = \int_{G/\Gamma\cap P} (\psi_1(x),f(x))\ dx ,$$

and, if $f(x) = \varphi(x)v(a(x))$, $\varphi \in {}^{o}L^2(M/\Gamma_M,\sigma,\chi)$, $v \in C_c^{\infty}(A)$,

$$((\psi_1,E_f))_{G/\Gamma} = \int_A v(a)e^{2\varrho(\log a)}da \int_{M/\Gamma_M} (\psi_{1,P}(ma),\varphi(m))dm = o .$$

Since v is arbitrary, this implies $\int_{M/\Gamma_M} (\psi_{1,P}(ma),\varphi(m))\,dm = 0$ for all φ. Hence by theorem 4, $\psi_1 = 0$. So $\alpha * \psi = 0$ for all $\alpha \in C_c(G)$ with ${}^k\alpha = \alpha$. Then $\psi = 0$.

For any $\vartheta \in \mathcal{E}_K$ (= set of all equivalence classes of finite-dimensional irreducible representations of K) we define the space ${}^0\mathcal{D}_{P|A}(\vartheta,\chi)$ as follows. Choose a representation (σ,V) belonging to the class ϑ. The space ${}^0\mathcal{D}_{P|A}(\vartheta,\chi)$ is spanned by the functions $x \to (v,f(x))$ with $v \in V$, $f \in {}^0\mathcal{D}_{P|A}(\sigma,\chi)$. For any finite subset F of $\mathcal{E}_K$ we define ${}^0\mathcal{D}_{P|A}(F,\chi) = \sum_{\vartheta \in F} {}^0\mathcal{D}_{P|A}(\vartheta,\chi)$.

As usually we denote by $d(\vartheta)$ the degree and by ξ_ϑ the character of ϑ and put $\alpha_\vartheta(k) = d(\vartheta)\overline{\xi_\vartheta(k)}$. Also, if $F \subset \mathcal{E}_K$ and F is finite, put $\alpha_F = \sum_{\vartheta \in F} \alpha_\vartheta$.

Fix a finite subset F of $\mathcal{E}_K$. Let V_F denote the set of all continuous functions β on K such that $\alpha_F * \beta = \beta$. Then V_F has finite dimension and we have a representation σ_F of K on V_F: $(\sigma_F(k_0)\beta)(k) = \beta(kk_0)$. Let f be any continuous function on G such that $\alpha_F * f = f$. Put $\beta_x(k) = f(kx)$ $(k \in K)$; $\beta_x \in V_F$ and $\sigma_F(k_0)\beta_x = \beta_{k_0 x}$. Write $\underline{\underline{f}}(x) = \beta_x$. Then $\underline{\underline{f}}$ is a σ_F-function on G. Moreover, if $f \in {}^0\mathcal{D}_{P|A}(F,\chi)$, then $\underline{\underline{f}} \in {}^0\mathcal{D}_{P|A}(\sigma_F,\chi)$ and the mapping $f \longmapsto \underline{\underline{f}}$ is a bijection of ${}^0\mathcal{D}_{P|A}(F,\chi)$ on ${}^0\mathcal{D}_{P|A}(\sigma_F,\chi)$.

<u>Lemma 43</u>. Let f be a C^∞ function on $G/U(\Gamma \cap P)$. Then $f \in {}^0\mathcal{D}_{P|A}(F,\chi)$ if and only if the following conditions hold.

1) $f = \alpha_F * f$.

2) $f(x) = 0$ unless $a(x)$ lies in some compact subset of A.

3) $\int_{M/\Gamma_M} |f(xm)|^2\,dm < \infty$.

4) Let P' be any cuspidal subgroup of G contained in P and different from P. Then $\int_{U'/U'\cap\Gamma} f(xu')\,du' = 0$.

5) $f(x;\zeta) = \chi(\zeta)f(x)$ for $\zeta \in \mathfrak{Z}_M$.

<u>Proof</u>. It is obvious that the conditions 1-5 are equivalent to the condition $\underline{\underline{f}} \in {}^0\mathcal{D}_{P|A}(\sigma_F,\chi)$.

<u>Corollary</u>. Let β be a left K-finite function in $C_c(G)$. Choose a finite subset F of $\mathcal{E}_K$ such that $\beta = \alpha_F * \beta$. Then $\beta * f \in {}^0\mathcal{D}_{P|A}(F,\chi)$ if $f \in {}^0\mathcal{D}_{P|A}(\vartheta,\chi)$, $\vartheta \in \mathcal{E}_K$.

Put

$$E_f(x) = \sum_{\Gamma/\Gamma\cap P} f(x\gamma) \qquad \text{for} \quad f \in {}^{o}\mathcal{D}_{P|A}(F,\chi)$$

and denote by $E_{P|A}(F,\chi)$ the space of all functions E_f with $f \in {}^{o}\mathcal{D}_{P|A}(F,\chi)$; if F consists of one element ϑ, write $E_{P|A}(F,\chi) = E_{P|A}(\vartheta,\chi)$. Let β and F be as in the corollary of Lemma 43. Then for $f \in {}^{o}\mathcal{D}_{P|A}(\vartheta,\chi)$, $\vartheta \in \mathcal{E}_K$, we have

$$\beta * E_f = E_{\beta * f} \in E_{P|A}(F,\chi) \ .$$

The spaces $E_{P|A}(\vartheta,\chi)$ are subspaces of $L^2(G/\Gamma)$. Define

$$E_{P|A}(\chi) = \mathrm{Cl}(\sum_{\vartheta\in\mathcal{E}_K} E_{P|A}(\vartheta,\chi)) \ .$$

Lemma 44. $E_{P|A}(\chi)$ is an invariant subspace of $L^2(G/\Gamma)$.

Proof. It is enough to prove that $f \in E_{P|A}(\vartheta,\chi)$, $\beta \in C_c^\infty(G)$ imply $\beta * f \in E_{P|A}(\chi)$. Since the left K-finite functions are dense in $C_c^\infty(G)$, it is enough to prove this when β is left K-finite, and then it follows from what we said above.

Note that $E_{G|1}(\vartheta,\chi) = {}^{o}L^2(G/\Gamma,\vartheta,\chi)$. These spaces generate ${}^{o}L^2(G/\Gamma)$ (Theorem 3). The spaces $E_{P|A}(\vartheta,\chi)$ (all (P,A), ϑ, χ) generate $L^2(G/\Gamma)$ as we prove in the next lemma.

Lemma 45. Assume the element g of $L^2(G/\Gamma)$ is orthogonal to $E_{P|A}(\chi)$ for all (P,A) and all $\chi \in \mathcal{H}_M$. Then $g = 0$.

Proof. Let $\mathcal{H}$ be the smallest closed invariant subspace of $L^2(G/\Gamma)$ which contains g. Then $\mathcal{H}$ is orthogonal to all $E_{P|A}(\chi)$. The K-finite vectors in $\mathcal{H}$ are zero as can be deduced from Lemma 42. Since these K-finite vectors are dense in $\mathcal{H}$, $\mathcal{H} = 0$.

§7. The adjoint of $c(s:\Lambda)$.

Let (P_1,A_1) and (P_2,A_2) be associate cuspidal pairs. Fix $s \in w(\mathfrak{a}_1,\mathfrak{a}_2)$. Let ${}^{o}\mathcal{D}_i(\sigma,\xi_i)$ $(i = 1,2)$ be as in Lemma 39. For $f_1 \in {}^{o}\mathcal{D}_1(\sigma,\xi_1)$ put

$$E_{f_1,s}(x) = \sum_{\Gamma(s)/\Gamma\cap P_1} f_1(x\gamma) \ .$$

$E_{f_1,s}$ is right-invariant under $\Gamma \cap P_2$.

Lemma 46. If $f_i \in {}^0\mathcal{D}_i(\sigma,\xi_i)$ $(i = 1,2)$, then

$$\int_{G/\Gamma\cap P_2} (f_2(x), E_{f_1,s}(x))dx = \int_{G/\Gamma\cap P_1} (E_{f_2,s^{-1}}(x), f_1(x))dx .$$

Proof. The integrals exist, because, in the notation of Lemma 26

$$\int_{G/\Gamma\cap P_2} |f_2(x)| \sum_{\Gamma/\Gamma\cap P_1} |f_1(x\gamma)| dx = \int_{G/\Gamma} E^*_{f_2} E^*_{f_1} dx < \infty .$$

$$\int_{G/\Gamma\cap P_2} \sum_{\Gamma(s)/\Gamma\cap P_1} (f_2(x), f_1(x\gamma))dx =$$

$$\sum_{\Gamma\cap P_2\backslash\Gamma(s)/\Gamma\cap P_1} \int_{G/\Gamma\cap P_2\cap {}^{\gamma}P_1} (f_2(x), f_1(x\gamma))dx =$$

$$\sum \int_{G/\Gamma\cap P_1\cap {}^{\gamma^{-1}}P_2} (f_2(y\gamma^{-1}), f_1(y))dy =$$

$$\sum_{\Gamma\cap P_1\backslash\Gamma(s^{-1})/\Gamma\cap P_2} \int_{G/\Gamma\cap P_1\cap {}^{\gamma}P_2} (f_2(y\gamma), f_1(y))dy$$

since $\Gamma(s)^{-1} = \Gamma(s^{-1})$.

Lemma 47. Under the assumptions of Lemma 46 we have

$$\int_{G/\Gamma\cap P_2} (f_2(x), E_{f_1,s}(x))dx = \int_{\mathfrak{a}_1^*} ((\hat{f}_2(-s\bar{\Lambda}), c(s:\Lambda)\hat{f}_1(\Lambda)))_{M_2/\Gamma_2} d\Lambda_I \quad (\Lambda_R \in (\mathfrak{a}_1^*)^-).$$

Proof. The proof is the same as that of Lemma 40 with the only difference that in all formulas we have in the present case a sum over $\Gamma(s)/\Gamma\cap P_1$ instead of $\Gamma/\Gamma \cap P_1$.

Lemma 48. Fix $s \in w(\mathfrak{a}_1, \mathfrak{a}_2)$ and let B denote the convex hull of $(\mathfrak{a}_1^*)^- \cup -s^{-1}(\mathfrak{a}_2^*)^-$. For any φ_1, φ_2 with $\varphi_i \in {}^0L^2(M_i/\Gamma_i, \sigma_i, \chi_i)$, $\chi_i \in \mathcal{H}_{M_i}$, we have

$$((\varphi_2, c(s:\Lambda)\varphi_1))_{M_2/\Gamma_2} = ((\varphi_2, c(s^{-1}:-s\bar{\Lambda})^*\varphi_1))_{M_2/\Gamma_2}$$

as holomorphic functions of Λ on $B + i\mathfrak{a}_1^*$.

Proof. Let $\varphi_i \in {}^0L^2(M_i/\Gamma_i,\sigma_i,\chi_i)$, $v_i \in C_c^\infty(A_i)$, $f_i(x) = \varphi_i(x)v_i(a_i(x))$ $(i = 1,2)$. Then $\hat{f}_i = \varphi_i\hat{v}_i$ and Lemma 47 gives

$$\int_{G/\Gamma\cap P_2} (f_2(x), E_{f_1,s}(x))\,dx = \int \overline{\hat{v}_2(-s\bar{\Lambda})}\ \hat{v}_1(\Lambda)((\varphi_2,c(s:\Lambda)\varphi_1))\ d\Lambda_I$$

and

$$\int_{G/\Gamma\cap P_1} (E_{f_2,s^{-1}}(x), f_1(x))\,dx = \int ((\hat{f}_2(M), c(s^{-1}:M)^*\hat{f}_1(-s^{-1}\bar{M})\ ))\ dM_I =$$

$$\int ((\hat{f}_2(-s\bar{\Lambda}), c(s^{-1}:-s\bar{\Lambda})^*\hat{f}_1(\Lambda)\ ))\ d\Lambda_I = \int \overline{\hat{v}_2(-s\bar{\Lambda})}\ \hat{v}_1(\Lambda)((\varphi_2,c(s^{-1}:-s\bar{\Lambda})^*\varphi_1))\ d\Lambda_I\ .$$

Put

$$F_1(\Lambda) = ((\varphi_2,c(s:\Lambda)\varphi_1))\ \hat{v}_1(\Lambda) \quad \text{for} \quad \Lambda \in (\mathfrak{a}^*_{1,c})^- ,$$

$$F_2(\Lambda) = ((\varphi_2,c(s^{-1}:-s\bar{\Lambda})^*\varphi_1))\ \hat{v}_1(\Lambda) \quad \text{for} \quad \Lambda \in -s^{-1}(\mathfrak{a}^*_{2,c})^- .$$

The two above formulas and Lemma 46 give

$$\int \overline{\hat{v}_2(-s\bar{\Lambda})}\ F_1(\Lambda)d\Lambda = \int \overline{\hat{v}_2(-s\bar{\Lambda})}\ F_2(\Lambda)d\Lambda ,$$

where in the left hand side integration is over $\mathrm{Re}(\Lambda) = \text{constant} \in (\mathfrak{a}^*_1)^-$ and in the right hand side over $\mathrm{Re}(\Lambda) = \text{constant} \in -s^{-1}(\mathfrak{a}^*_2)^-$. Observe that F_1 and F_2 are Schwartz functions.

$$\int \overline{\hat{v}_2(-s\bar{\Lambda})}\ F_j(\Lambda)d\Lambda = \int_{A_2} \overline{v_2(a)}\ da \int F_j(\Lambda)e^{(s\Lambda+\varrho_2)(\log a)}d\Lambda \quad (j = 1,2) ,$$

hence, since v_2 is arbitrary,

$$\int F_1(\Lambda)e^{s\Lambda(\log a)}d\Lambda = \int F_2(\Lambda)e^{s\Lambda(\log a)}d\Lambda \qquad \text{for} \quad a \in A_2 .$$

Put for $a \in A_1$

$$\int_{\mathfrak{a}^*_1} F_1(\lambda_1+i\mu)e^{(\lambda_1+i\mu)(\log a)}d\mu = \int_{\mathfrak{a}^*_1} F_2(\lambda_2+i\mu)e^{(\lambda_2+i\mu)(\log a)}d\mu = f(a) .$$

Here $\lambda_1 \in (\mathfrak{a}^*_1)^-$, $\lambda_2 \in -s^{-1}(\mathfrak{a}^*_2)^-$; f is independent of λ_1, λ_2 , and we have

$$F_j(\lambda_j+i\mu) = \int_{A_1} f(a)\ e^{-(\lambda_j+i\mu)(\log a)}da \quad (j = 1,2).$$

Let ω_1^* and ω_2^* be compact subsets of $(\mathfrak{a}_1^*)^-$ and $-s^{-1}(\mathfrak{a}_2^*)^-$ and ω^* the convex hull of $\omega_1^* \cup \omega_2^*$. For any number N there is a constant b such that

$$(1+\|\log a\|)^N |f(a)|\, a^{-\lambda} \leq b \qquad \text{for } a \in A_1,\ \lambda \in \omega_1^* \cup \omega_2^* .$$

The same inequality holds then for $\lambda \in \omega^*$. So the integral $\int_{A_1} |f(a)| e^{-\lambda(\log a)} da$ converges uniformly for $\lambda \in \omega^*$. Hence the function F defined by

$$F(\Lambda) = \int_{A_1} f(a) e^{-\Lambda(\log a)} da$$

is holomorphic on $B + i\mathfrak{a}_1^*$. Moreover $F = F_1$ on $(\mathfrak{a}_{1,c}^*)^-$ and $F = F_2$ on $-s^{-1}(\mathfrak{a}_{2,c}^*)^-$. Since v_1 is arbitrary, this implies that $((\varphi_2, c(s:\Lambda)\varphi_1))$ and $((\varphi_2, c(s^{-1}:-s\bar{\Lambda})^*\varphi_1))$ extend to one and the same holomorphic function on $B + i\mathfrak{a}_1^*$.

§8. Automorphic forms of type q .

Let q be an integral number and σ a finite-dimensional representation of K. We define $\mathcal{A}_q(G/\Gamma,\sigma)$ to be the space of those elements f of $\mathcal{A}(G/\Gamma,\sigma)$ for which $f_P \sim 0$ for any cuspidal subgroup P of G with rank $P \neq q$ (the relation $f_P \sim 0$ was defined in §4). If $q < 0$ or $q > \mathrm{rank}_{\mathbb{Q}}(G)$, then $\mathcal{A}_q(G/\Gamma,\sigma) = 0$ (Theorem 4).

Lemma 49. Let f be an element of $\mathcal{A}_q(G/\Gamma,\sigma)$ and $P = MAU$ a cuspidal subgroup. Fix $a \in A$ and put $g(m) = f_P(ma)$ $(m \in M)$. Then $g \in \mathcal{A}_{q-\mathrm{rank}\,P}(M/\Gamma_M,\sigma_M)$.

Proof. Let $^*P$ be a cuspidal subgroup of M and let P' be the corresponding cuspidal subgroup of G contained in P. Then

$$\int_{^*M/\Gamma_{^*M}} (\varphi(^*m), g_{^*P}(^*m^*a))\, d^*m = \int_{M'/\Gamma_{M'}} (\varphi(m'), f_{P'}(m'^*aa))\, dm'$$

for $\varphi \in {}^0L^2(M'/\Gamma_{M'},\sigma_{M'},\chi)$, $\chi \in \mathcal{H}_{M'}$ and $^*a \in {}^*A$ (cf. the proof of Theorem 4). By assumption the second integral is zero unless rank $P' = q$, i.e. unless rank $^*P =$ $= q -$ rank P.

Corollary 1. If $f \in \mathcal{A}_q(G/\Gamma,\sigma)$ and rank $P > q$, then $f_P = 0$.

Corollary 2. $\mathcal{A}_0(G/\Gamma,\sigma) = {}^0\mathcal{A}(G/\Gamma,\sigma)$.

Lemma 50. Suppose $g \in \mathcal{A}(G/\Gamma,\sigma)$. Then $g \in \mathcal{A}_q(G/\Gamma,\sigma)$ if and only if the following condition holds. Let (P,A) be a cuspidal pair with rank $P \neq q$. Then for any $\chi \in \mathcal{H}_M$ we have

$$\int_{G/\Gamma} (g(x),E_f(x))\,dx = 0$$

for all $f \in {}^0\mathcal{D}_{P|A}(\sigma,\chi)$.

Proof. If $f(x) = \varphi(x)v(a(x))$, then

$$\int_{G/\Gamma} (g(x),E_f(x))\,dx = \int_A e^{2\varrho(\log a)} v(a)\,da \int_{M/\Gamma_M} (g_P(am),\varphi(m))\,dm$$

(Lemma 27).

Corollary. Let $\alpha \in I_c^\infty(G)$ and $g \in \mathcal{A}_q(G/\Gamma,\sigma)$. Then $\alpha * g \in \mathcal{A}_q(G/\Gamma,\sigma)$.

Proof. $\int(\alpha * g,E_f)\,dx = \int(g,\tilde{\alpha} * E_f)\,dx = \int(g,E_{\tilde{\alpha} * f})\,dx$ by Lemma 41.

Lemma 51. Let $\{P_1,\ldots,P_r\}$ be a complete set of representatives for the Γ-conjugacy classes of cuspidal subgroups of G of rank q. Let $f \in \mathcal{A}_q(G/\Gamma,\sigma)$.

a) If $f_{P_i} = 0$ for $1 \le i \le r$, then $f = 0$.

b) If f_{P_i} is $\mathfrak{Z}$-finite for $1 \le i \le r$, then f is $\mathfrak{Z}$-finite.

Proof. Since $f_{\gamma_P}(x) = f_P(x\gamma)$ for $\gamma \in \Gamma$, a) follows from Theorem 4. If the functions f_{P_i} are $\mathfrak{Z}$-finite, we can choose a function α in $I_c^\infty(G)$ such that $f_{P_i} = \alpha * f_{P_i}$ for all i. By the corollary of Lemma 50, $\alpha * f \in \mathcal{A}_q(G/\Gamma,\sigma)$, so $g = f - \alpha * f \in \mathcal{A}_q(G/\Gamma,\sigma)$. On the other hand $g_{P_i} = f_{P_i} - \alpha * f_{P_i} = 0$, so by a) we have $g = 0$, that is $f = \alpha * f$.

There exists an ideal $\mathfrak{U}$ of finite codimension in $\mathfrak{Z}$ such that $uf_{P_i} = 0$ for $u \in \mathfrak{U}$, $1 \le i \le r$. Now $uf = u(\alpha * f) = (u\alpha) * f$, hence $uf \in \mathcal{A}_q(G/\Gamma,\sigma)$ (Corollary of Lemma 50), and $(uf)_{P_i} = u\, f_{P_i} = 0$ if $u \in \mathfrak{U}$, so $uf = 0$ for $u \in \mathfrak{U}$.

CHAPTER III

§0. Introduction.

Let f be a $\mathfrak{Z}$-finite element of $\mathcal{A}_q(G/\Gamma,\sigma)$. We want to estimate f in terms of the f_P, P running through the set of cuspidal subgroups of G with rank q (or through a set of representatives modulo Γ of those cuspidal groups). The idea is the following. Fix $P = MAU$ with rank q. The restriction of f_P on MA is $\mathfrak{Z}_M\mathfrak{a}$-finite (because $\mathfrak{Z}_M\mathfrak{a}$ is a finite module over $\mu'(\mathfrak{Z})$). Moreover, for each $a \in A$ the function $m \to f_P(ma)$ lies in ${}^{\circ}\mathcal{A}(M/\Gamma_M,\sigma_M)$ (Lemma 49 and its Corollary 2). So $f_P|_{MA}$ is of the form

$$(*) \qquad f_P(ma) = \sum \varphi_i(m)\, p_i(\log a)\, e^{\Lambda_i(\log a)}$$

where i runs through a finite set of indices, φ_i is a $\mathfrak{Z}_M$-finite element of ${}^{\circ}\mathcal{A}(M/\Gamma_M,\sigma_M)$, p_i resp. Λ_i is a polynomial resp. a linear function on $\mathfrak{a}_c^*$. Roughly speaking, what Langlands does is this: let the φ_i be limited to belong to a finite-dimensional $\mathfrak{Z}_M$-stable subspace of ${}^{\circ}\mathcal{A}(M/\Gamma_M,\sigma_M)$, the p_i to have bounded degree and the Λ_i to vary in a compact set; then find an estimate for f. However, Langlands' lemma is not quite sufficient for our purpose. The reason is, that, when the eigenvalues Λ_i are given and they are not all different, then the decomposition $(*)$ is not unique. So we shall proceed somewhat differently in the case $q = 1$.

§1. Statement of Theorem 6.

Fix a cuspidal subgroup $P = MAU$. Let Φ be a finite-dimensional $\mathfrak{Z}_M$-stable subspace of ${}^{\circ}\mathcal{A}(M/\Gamma_M,\sigma_M)$ and χ a representation of A on a finite-dimensional space B^*. If V is as usual the space of the representation σ, the space $V \otimes_C B^*$ can be regarded as a (K,A)-bimodule $(k \longmapsto \sigma(k) \otimes 1,\ a \longmapsto 1 \otimes \chi(a))$. Let h be an element of $\Phi \otimes B^*$. We regard it as a function $M \to V \otimes B^*$ and define

$$h_\chi(k\,mau) = \sigma(k)h(m)\chi(a) \quad .$$

If $\alpha \in C_c(G)$ such that ${}^k\alpha = \alpha$ for $k \in K$, and $h \in \Phi \otimes B^*$, then

$$(\alpha * h_\chi)(x) = \int_M \tau_\alpha(xm^{-1})h(m)\,dm \qquad (x \in G),$$

where

$$\tau_\alpha(x) = \int\int\int_{K\times A\times U} \alpha(k^{-1}xa^{-1}u)(\sigma(k) \otimes \chi(a))\,dk\,da\,du \quad (x \in G).$$

From this one sees that, if Φ is invariant under convolution with the functions $\gamma \in C_c(M) \otimes \mathrm{End}(V)$ which have the property that $\gamma(km) = \sigma_M(k)\gamma(m)$ for $k \in K_M$, $m \in M$, then the restriction h' of $\alpha * h_\chi$ on M lies in $\Phi \otimes B^*$ and $\alpha * h_\chi = (h')_\chi$. Under this assumption on Φ we get in particular a representation $\alpha \longmapsto \pi(\chi:\alpha)$ of $I_c^\infty(G)$ on $\Phi \otimes B^*$; it is defined by

$$\alpha * h_\chi = (\pi(\chi:\alpha)h)_\chi \qquad (h \in \Phi \otimes B^*).$$

The condition on Φ is certainly fulfilled if Φ is a space ${}^\circ\mathcal{A}(M/\Gamma_M,\sigma_M,\xi)$ with $\xi \subset \mathcal{H}_M$, ξ finite.

Assume now that the A-module B^* is the dual of an A-module B where the underlying vector space of B is a finite-dimensional subspace of $\mathfrak{O}$ containing 1. We write χ for both representations, so $(b^*\chi(a))(b) = b^*(\chi(a)b)$ for $a \in A$, $b \in B$, $b^* \in B^*$. Choose a base $b_1,\dots,b_N$ for B, let $b_1^*,\dots,b_N^*$ be the dual base for B^* and let b_i' be the image of b_i under the automorphism of $\mathfrak{O}$ defined by $H \longmapsto H + \varrho(H)$ for $H \in \mathfrak{a}$ ($b_i' = e^{-\varrho}b_i \circ e^{\varrho}$ as differential operators). For any $\mathfrak{Z}$-finite function in $\mathcal{A}(G/\Gamma,\sigma)$ we put

$$f^P(x) = \sum_{1\le i\le N} f_P(x;b_i') \otimes b_i^* \qquad (x \in G).$$

We assume that B is supplied with some Hilbert-space structure such that $|1| = 1$. Considering the elements of $V \otimes B^*$ as linear mappings of B into V we have then $|f_P(x)| = |f^P(x)(1)| \le |f^P(x)|$.

Let $P_1,\dots,P_r$ be a complete set of representatives for the Γ-conjugacy classes of cuspidal subgroups of G of rank 1. Fix for each i a split component A_i of P_i. Let B_i be a finite-dimensional subspace of $\mathfrak{O}_i$ containing 1. We choose a Hilbert-space structure on B_i such that $|1| = 1$.

Let D be a topological space and suppose that for any $z \in D$ we are given representations $\chi_i(z)$ of A_i on B_i $(1 \le i \le r)$. We assume that for fixed $a \in A_i$ the mapping $z \longmapsto \chi_i(z:a)$ is continuous.

Let for each i, Φ_i be a $\mathfrak{Z}_{M_i}$-stable finite-dimensional subspace of ${}^o\mathcal{A}(M_i/\Gamma_i,\sigma_i)$ $(\Gamma_i = \Gamma_{M_i}, \sigma_i = \sigma_{M_i})$ and put

$$\underline{\underline{\Phi}} = \prod_{1 \le i \le r} \Phi_i \otimes B_i^*$$

<u>Definition</u>. $L(z) = \mathcal{A}_1(G/\Gamma,\sigma,\underline{\underline{\Phi}},z)$ is the space of all $\mathfrak{Z}$-finite functions f in $\mathcal{A}_1(G/\Gamma,\sigma)$ with the property that there exists an element $h = (h_i)$ of $\underline{\underline{\Phi}}$ such that $f^{P_i} = (h_i)_{\chi_i(z)}$ for $1 \le i \le r$.

<u>Definition</u>. $\underline{\underline{\Phi}}(z)$ is the space of all $h \in \underline{\underline{\Phi}}$ for which there exists an element f of $L(z)$ such that $f^{P_i} = (h_i)_{\chi_i(z)}$ for $1 \le i \le r$.

For given $h \in \underline{\underline{\Phi}}(z)$ the element f of $L(z)$ in the second definition is unique, because $f^{P_i} = o$ for $1 \le i \le r$ implies $f_{P_i} = o$ for $1 \le i \le r$, which implies $f = o$ (Theorem 4); write $f = f(h:z)$. The mapping $h \longmapsto f(h:z)$ is a bijection of $\underline{\underline{\Phi}}(z)$ on $L(z)$.

We shall use on the finite-dimensional vector space $\underline{\underline{\Phi}}$ the norm μ defined by

$$\mu(h) = \max_{1 \le i \le r} \mu(h_i) \ , \qquad \mu(h_i) = \sup_{m \in M_i} |h_i(m)|$$

for $h = (h_i) \in \underline{\underline{\Phi}}$.

<u>Theorem 6</u>. Let (P_o,A_o) be a mincuspidal pair. Fix $\xi_i \in G_{\mathbb{Q}}$ such that ${}^{\xi_i}(P_i,A_i) \succ (P_o,A_o)$. Then for any Siegel domain $\mathfrak{S}$ in G with respect to (P_o,A_o) and any compact set ω in D there exists a number c such that

$$|f(h:z:x)| \le c\,\mu(h) \sum_{1 \le i \le r} |\chi_i(z:\exp(H_i(x\xi_i)))|$$

for all $z \in \omega$, $h \in \underline{\underline{\Phi}}(z)$, $x \in \mathfrak{S}$.

Remark. Theorem 6 suffices to estimate $f(h:z)$ on G: one applies the theorem to each element of a complete set of representatives of the mincuspidal pairs of G modulo conjugation under Γ, the Siegel domains being chosen in such a way that there union covers G modulo Γ.

§2. Proof of Theorem 6.

Fix $F \subset \Sigma^o = \Sigma^o(P_o|A_o)$ with rank $F = 1$, i.e. F contains all simple roots but one. Put $(P,A) = (P_o,A_o)_F$. There is an element γ in Γ and a unique i such that $P = {}^{\gamma}P_i$ and there is an element u in $U_{i,\mathbb{Q}}$ such that $(P,A) = {}^{\gamma u}(P_i,A_i)$. Both ${}^{\gamma u}(P_i,A_i)$ and ${}^{\xi_i}(P_i,A_i)$ dominate (P_o,A_o), so they are equal. Hence $\gamma u = \xi_i m_i a_i$ with $m_i \in M_i$, $a_i \in A_i$. For any continuous function f on G/Γ we have

$$f_P(x) = f_{P_i}(x\gamma) = f_{P_i}(x\gamma u) = f_{P_i}(x\xi_i m_i a_i) .$$

For $f = f(h:z)$, $h \in \underline{\Phi}(z)$, we have, immediately from the definitions,

$$|f_{P_i}(x)| \leq |f^{P_i}(x)| \leq \mu(h_i)|\chi_i(z:a_i(x))| ,$$

where $a_i(x) = \exp(H_i(x))$. Since $H_i(x\xi_i m_i a_i) = H_i(x\xi_i) + \log a_i$, we get

$$|f_P(x)| \leq \mu(h_i)|\chi_i(z:\exp(H_i(x\xi_i)))||\chi_i(z:a_i)| .$$

If z stays in ω, then $|\chi_i(z:a_i)|$ is bounded, so we have proved:

Lemma 52. Put $P = P_{o,F}$ where rank $F = 1$. Let P be conjugate to P_i under Γ. Then there exists a constant c such that

$$|f_P(x)| \leq c\,\mu(h_i)|\chi_i(z:a_i(x\xi_i))|$$

for $z \in \omega$, $h \in \underline{\Phi}(z)$, $f = f(h:z)$, $x \in G$.

Fix a set of representatives Ξ for $P_{o,\mathbb{Q}}\backslash G_{\mathbb{Q}}/\Gamma$. Put $\Gamma_\infty = U_o \cap \bigcap_{\xi\in\Xi} {}^{\xi}\Gamma$. For $f \in \mathcal{A}(G/\Gamma,\sigma)$, $\xi \in \Xi$, $F \subset \Sigma^o$ put

$$f_\xi(x) = f(x\xi) \ , \quad f_{\xi,F}(x) = \int_{U_{o,F}/U_{o,F}\cap\Gamma_\infty} f_\xi(xu)\,du \ .$$

Applying Lemma 52 to ${}^{\xi^{-1}}(P_o,A_o)$ instead of (P_o,A_o) and observing that

$$f_{\xi,F}(x) = (U_{o,F}\cap {}^{\xi}\Gamma : U_{o,F}\cap\Gamma_\infty) \int_{{}^{\xi^{-1}}U_{o,F}/{}^{\xi^{-1}}U_{o,F}\cap\Gamma} f(x\xi u)\,du \ ,$$

we find the following.

Corollary. Let $F \subset \Sigma^o$ with rank $F = 1$ and $\xi \in \Xi$. If ${}^{\xi^{-1}}P_{o,F}$ is conjugate to P_i under Γ , then there exists a constant c such that

$$|f_{\xi,F}(x)| \leq c\,\mu(h_i)|\chi_i(z:a_i(x\xi_i))|$$

for $z \in \omega$, $h \in \underline{\Phi}(z)$, $f = f(h:z)$, $x \in G$.

By Corollary 1 of Lemma 49, $f_{\xi,F} = o$ if rank $F > 1$. So we get:

Lemma 53. There is a constant c such that

$$\sum_{F\neq\Sigma^o} |f_{\xi,F}(x)| \leq c\,\mu(h) \sum_{1\leq i\leq r} |\chi_i(z:a_i(x\xi_i))|$$

for $z \in \omega$, $h \in \underline{\Phi}(z)$, $f = f(h:z)$, $\xi \in \Xi$, $x \in G$.

In the notation of Lemma 11 we have

$$f_\xi = {}^o f_\xi - \sum_{F\neq\Sigma^o} (-1)^{\operatorname{rank} F} f_{\xi,F} \ .$$

So we have to estimate ${}^o f_\xi$.

Lemma 54. There are constants c_1,c_2 with $o < c_1 \leq c_2$ and $\lambda \in \mathfrak{a}_o^*$ such that

$$c_1\, e^{\lambda(H_o(x))} \leq \sum_{1\leq i\leq r} |\chi_i(z:a_i(x\xi_i))| \leq c_2\, e^{-\lambda(H_o(x))}$$

for $x \in \mathfrak{F}$, $z \in \omega$.

Proof. If $x = kap$, $k \in K$, $a \in A_o$, $p \in M_oU_o$, then $x\xi_i = k\xi_i\, {}^{\xi_i^{-1}}a\, {}^{\xi_i^{-1}}p$ and

$a_i(x\xi_i) = a_i(k\xi_i)a_i(\xi_i^{-1}a)$, hence $\chi_i(z:a_i(x\xi_i)) = \chi_i(z:a_i(k\xi_i))\chi_i(z:a_i(\xi_i^{-1}a))$ and we can choose c_1', c_2' with $0 < c_1' \leq c_2'$ such that

$$c_1' \sum |\chi_i(z:a_i(\xi_i^{-1}a))| \leq \sum |\chi_i(z:a_i(x\xi_i))| \leq c_2' \sum |\chi_i(z:a_i(\xi_i^{-1}a))|$$

for all $z \in \omega$, $x \in G$. Here $a = a_o(x) = \exp(H_o(x))$. It is trivial that we can choose $\lambda \in \mathfrak{a}_o^*$ and constants $c_1'', c_2'' > 0$ such that

$$c_1'' e^{\lambda(\log a)} \leq \sum |\chi_i(z:a_i(\xi_i^{-1}a))| \leq c_2'' e^{-\lambda(\log a)}$$

for $a \in A_{o,t}$, $z \in \omega$. This gives the lemma.

Lemma 55. For any $z \in \omega$ there is a number $c(z)$ such that

$$|{}^o f_\xi(x)| \leq c(z)\ \mu(h) \sum_{1\leq i\leq r} |\chi_i(z:a_i(x\xi_i))|$$

for $h \in \underline{\Phi}(z)$, $f = f(h:z)$, $\xi \in \Xi$, $x \in \mathfrak{Y}$.

Proof. Let λ be as in Lemma 54. Consider for $h \in \underline{\Phi}(z)$ and $f = f(h:z)$,

$$\nu(h) = \sup_{x\in\mathfrak{Y},\xi\in\Xi} |{}^o f_\xi(x)|\ e^{-\lambda(H_o(x))}.$$

By the corollary of Lemma 10, $\nu(h)$ is finite, hence ν is a seminorm on the finite-dimensional space $\underline{\Phi}(z)$, so there is a constant $c(z)$ such that $\nu(h) \leq c(z)\ \mu(h)$ for all $h \in \underline{\Phi}(z)$. Using Lemma 54 we get Lemma 55.

Lemma 56. For any $z \in \omega$ there exists a number $c(z)$ such that

$$|f(h:z:x\xi)| \leq c(z)\ \mu(h) \sum_{1\leq i\leq r} |\chi_i(z:a_i(x\xi_i))|$$

for $h \in \underline{\Phi}(z)$, $x \in \mathfrak{Y}$, $\xi \in \Xi$.

Proof. This follows from Lemmas 11, 53 and 55.

Theorem 6 will be proved when we show that it is possible to choose $c(z)$ in such a way that it is bounded on ω. When we enlarge the Φ_i and $\mathfrak{Y}$, the statement of the theorem becomes stronger. So we may assume that:

1) For any $z \in \omega$ and any i we have a representation $\alpha \longmapsto \pi_i(z:\alpha)$ of $I_c^\infty(G)$ on $\Phi_i \otimes B_i^*$ such that

$$\alpha * (h_i)_{\chi_i(z)} = (\pi_i(z:\alpha)h_i)_{\chi_i(z)} \qquad (h_i \in \Phi_i \otimes B_i^*)$$

[see §1].

2) $\mathcal{J}U_o = \mathcal{J}\Gamma_\infty$.

3) $G = {}^o\mathcal{J}\Xi\Gamma$ where ${}^o\mathcal{J}$ is the interior of $\mathcal{J}$.

The representations $\pi_i(z)$ $(1 \leq i \leq r)$ define a representation $\pi(z)$ of $I_c^\infty(G)$ on $\underline{\underline{\Phi}}$: $\pi(z:\alpha)h = (\pi_i(z:\alpha)h_i)$ if $h = (h_i) \in \underline{\underline{\Phi}}$. The subspace $\underline{\underline{\Phi}}(z)$ of $\underline{\underline{\Phi}}$ is invariant under $\pi(z:\alpha)$ (Corollary of Lemma 50).

Let, for $z \in \omega$, $c(z)$ denote the <u>smallest</u> number ≥ 0 for which the inequality in Lemma 56 holds for all $h \in \underline{\underline{\Phi}}(z)$, $x \in \mathcal{J}$, $\xi \in \Xi$.

<u>Lemma 57</u>. Let z and h vary in ω and $\underline{\underline{\Phi}}$ respectively, in such a way that $h \in \underline{\underline{\Phi}}(z)$, that $\mu(h)$ remains bounded and that $c(z) > 0$. Put $f = c(z)^{-1}f(h:z)$. Then f remains bounded on every compact subset of G .

<u>Proof</u>. Let C be a compact subset of G . From the assumption that $G = {}^o\mathcal{J}\Xi\Gamma$ it follows trivially that C is contained in some set of the form $C_o\Xi\Gamma$ with $C_o \subset \mathcal{J}$ and C_o compact. For $x \in C_o$, $\xi \in \Xi$, $\gamma \in \Gamma$ we have by Lemmas 56 and 54

$$|f(x\xi\gamma)| = c(z)^{-1}|f(h:z:x\xi)| \leq c_2\mu(h)e^{-\lambda(H_o(x))} .$$

The last expression is bounded, by assumption.

Suppose now that $\sup_{z\in\omega} c(z) = \infty$.

Then we can choose a convergent sequence (z_n) in ω such that $c(z_n)$ tends to infinity and that $c(z_n) > 0$ for all n . Let $\lim z_n = z_o$. From $c(z_n) > 0$ it follows that $\underline{\underline{\Phi}}(z_n) \neq 0$. Put

$$\nu_n(x) = \sum_{1\leq i\leq r} |\chi_i(z_n:a_i(x\xi_i))| \qquad (x \in G).$$

Now we choose for each n an element h_n of $\underline{\underline{\Phi}}(z_n)$ with $\mu(h_n) = 1$ such that, if we put $f_n = c(z_n)^{-1} \; f(h_n:z_n)$, then

$$\sup_{x\in\mathfrak{Y},\xi\in\Xi} |f_n(x\xi)| \; \nu_n(x)^{-1} > \frac{1}{2} .$$

We may also assume, after taking a subsequence, that h_n converges to h_o in $\underline{\underline{\Phi}}$.

Since the functions $(h_i)_{\chi_i(z_o)}$ $(h_i \in \Phi_i \otimes B_i^*)$ form a finite-dimensional space which is invariant under K and $\mathfrak{Z}$, we can choose $\alpha \in I_c^\infty(G)$ such that $\pi(z_o:\alpha) = 1$. Then $\lim \pi(z_n:\alpha) = 1$, so $\pi(z_n:\alpha)$ is bijective for almost all n ; suppose it is so for all n . Since $\underline{\underline{\Phi}}(z_n)$ is stable under $\pi(z_n:\alpha)$, the element $h_n' = \pi(z_n:\alpha)^{-1}h_n$ of $\underline{\underline{\Phi}}$ lies in $\underline{\underline{\Phi}}(z_n)$. We have

$$\lim h_n' = \lim \pi(z_n:\alpha)^{-1}h_n = \pi(z_o:\alpha)^{-1}h_o = h_o ;$$

in particular $\mu(h_n')$ is bounded, assume $\mu(h_n') \leq 2$. Put

$$f_n' = c(z_n)^{-1} \; f(h_n':z_n) .$$

Then $\alpha * f_n' = f_n$, because $\alpha * f(h_n':z_n) = f(\pi(z_n:\alpha)h_n':z_n) = f(h_n:z_n)$. By Lemma 57 the sequence f_n' remains bounded on every compact subset of G . Hence by taking a suitable subsequence we may assume that $\alpha * f_n'$ converges uniformly on every compact set (see Lemma 59 at the end of this paragraph). Define $f(x) = \lim f_n(x)$. Obviously ,
$f \in C(G/\Gamma,\sigma)$. The definition of f_n and Lemmas 56 and 54 give

$$|f_n(x\xi)| \leq \nu_n(x) \leq c_2 \, e^{-\lambda(H_o(x))} \qquad (x \in \mathfrak{Y}, \; \xi \in \Xi) ,$$

hence $|f(x\xi)| \leq c_2 \, e^{-\lambda(H_o(x))}$, so that $f \in \mathcal{A}(G/\Gamma,\sigma)$. In the same way, from Lemmas 53 and 54 we find

$$|f_{n,\xi,F}(x)| \leq c \, c_2 \, c(z_n)^{-1} \, e^{-\lambda(H_o(x))} \qquad (x \in \mathfrak{Y}, \; \xi \in \Xi),$$

if rank $F > o$, and from this

$$f_{\xi,F}(x) = \lim f_{n,\xi,F}(x) = o \qquad (x \in \mathfrak{Y}, \; \xi \in \Xi).$$

So $f \in {}^0\mathcal{A}(G/\Gamma,\sigma)$. Using the above estimate for $f_n(x\xi)$ one proves that

$$\int_{G/\Gamma} (\varphi(x),f(x))dx = \lim \int_{G/\Gamma} (\varphi(x),f_n(x))dx$$

for $\varphi \in {}^0L^2(G/\Gamma,\sigma,\chi)$, $\chi \in \mathcal{H}$; since $f_n \in \mathcal{A}_1(G/\Gamma,\sigma)$ the integrals on the right hand side are zero, hence $f = o$ (Theorem 4). Thus we have proved that f_n converges uniformly to 0 on every compact set.

Let ν'_{g,λ_o} $(g \in \mathcal{G},\ \lambda_o \in \mathfrak{a}_o^*)$ be defined as in Chapter I,§4, i.e.

$$\nu'_{g,\lambda_o}(\varphi) = \sup_{x\in\mathcal{F},\xi\in\Xi} |\varphi(g.x\xi)|\ e^{\lambda_o(H_o(x))} \qquad (\varphi \in C^\infty(G/\Gamma)).$$

<u>Lemma 58</u>. Let α and λ be as above. There exists $\lambda_1 \in \mathfrak{a}_o^*$ with the following property. For any $g \in \mathcal{G}$ there exists a number c_g such that

$$\nu'_{g,\lambda_1}(\alpha*\varphi) \leq c_g\ \nu'_{1,\lambda}(\varphi) \qquad (\varphi \in C^\infty(G/\Gamma)).$$

<u>Proof</u>. Define for any real number r, $\eta_r(\varphi) = \sup_{x\in G} |\varphi(x)|\,\|x\|^{-r}$ $(\leq \infty)$. Choose numbers b_1 and r_1 such that $\eta_{r_1}(\varphi) \leq b_1\nu'_{1,\lambda}(\varphi)$ for all $\varphi \in C^\infty(G/\Gamma)$ (Lemmas 4 and 5). Then, for $\beta \in C_c(G)$, $\varphi \in C^\infty(G/\Gamma)$,

$$|(\beta*\varphi)(x)| \leq \eta_{r_1}(\varphi) \int |\beta(y)|\,\|y^{-1}x\|^{r_1}\,dy \leq$$

$$\eta_{r_1}(\varphi)\,\|x\|^{r_1} \int |\beta(y)|\,\|y^{-1}\|^{r_1}\,dy = c(\beta)\eta_{r_1}(\varphi)\,\|x\|^{r_1},$$

so $\eta_{r_1}(\beta*\varphi) \leq c(\beta)\eta_{r_1}(\varphi) \leq c(\beta)b_1\ \nu'_{1,\lambda}(\varphi)$. On the other hand, $\eta_{r_1}(\psi) \geq b_2^{-1}\nu'_{1,\lambda_1}(\psi)$ for some real number b_2 and some $\lambda_1 \in \mathfrak{a}_o^*$. Hence $\nu'_{1,\lambda_1}(\beta*\varphi) \leq b_1b_2c(\beta)\nu'_{1,\lambda}(\varphi)$. In particular $\nu'_{g,\lambda_1}(\alpha*\varphi) = \nu'_{1,\lambda_1}(g'\alpha*\varphi) \leq c_g\ \nu'_{1,\lambda}(\varphi)$.

Choose λ_1 with the property stated in Lemma 58. Given any $\lambda' \in \mathfrak{a}_o^*$, we can choose $g_1,\dots,g_N \in \mathcal{G}$ such that

$$(*) \qquad \sup_{x\in\mathcal{F},\xi\in\Xi} |{}^0\psi_\xi(x)|\ e^{\lambda'(H_o(x))} \leq \sum_{1\leq i\leq N} \nu'_{g_i,\lambda_1}(\psi)$$

for all $\psi \in \mathcal{A}^\infty(G/\Gamma,\lambda_1)$ (Lemma 10). Take $\lambda' = -\lambda - \varrho_o$. Since

$$|f_n'(x\xi)| \leq \mu(h_n')\, \nu_n(x) \leq 2\, c_2\, e^{-\lambda(H_o(x))} \qquad \text{for} \quad x \in \mathfrak{F},\ \xi \in \Xi ,$$

$\nu'_{1,\lambda}(f_n') \leq 2c_2$ and $\nu'_{g,\lambda_1}(f_n) = \nu'_{g,\lambda_1}(\alpha * f_n') \leq 2c_g c_2$. So from the inequality (*) we get

$$|{}^o f_{n,\xi}(x)| \leq c_3\, e^{(\lambda+\varrho_o)(H_o(x))} \qquad (x \in \mathfrak{F},\ \xi \in \Xi,\ n \geq 1).$$

On the other hand, by Lemma 53 we have

$$\sum_{F \neq \Sigma^o} |f_{n,\xi,F}(x)| \leq c\, c(z_n)^{-1}\, \nu_n(x) .$$

Hence, by Lemmas 11 and 54,

$$|f_n(x\xi)|\,\nu_n(x)^{-1} \leq c_4\, e^{\varrho_o(H_o(x))} + c_5\, c(z_n)^{-1} \qquad (x \in \mathfrak{F},\ \xi \in \Xi,\ n \geq 1).$$

And, by Lemma 54,

$$|f_n(x\xi)|\,\nu_n(x)^{-1} \leq c_1^{-1}\, e^{-\lambda(H_o(x))}\, |f_n(x\xi)| \qquad (x \in \mathfrak{F},\ \xi \in \Xi,\ n \geq 1).$$

Since $e^{\varrho_o(H_o(x))}$ tends to zero on the complements of compact subsets of $\mathfrak{F}$, one sees from these two inequalities that $f_n(x\xi)\nu_n(x)^{-1}$ tends to zero uniformly for $x \in \mathfrak{F}$, $\xi \in \Xi$. This is in contradiction with the assumption that $\sup_{x\in\mathfrak{F},\xi\in\Xi} |f_n(x\xi)|\nu_n(x)^{-1} > \frac{1}{2}$. Theorem 6 is proved.

During the proof we have used the following little lemma, which we shall prove in order to be complete.

Lemma 59. Let φ_n be a sequence in $C(G/\Gamma)$. Assume that φ_n remains bounded on every compact subset of G. Let $\alpha \in C_c(G)$. Then by selecting a subsequence we can arrange that $\alpha * \varphi_n$ converges uniformly on every compact subset of G.

Proof. Choose a sequence (C_k) of compact subsets of G/Γ such that C_k is contained in the interior of C_{k+1} and that $\cup C_k = G/\Gamma$, and a sequence of functions $\beta_k \in C(G/\Gamma)$ such that $\beta_k = 1$ on C_k, $\beta_k = 0$ outside C_{k+1} and $0 \leq \beta_k \leq 1$ everywhere. Then $\|\beta_k\varphi_n\|^2_{G/\Gamma} \leq \int_{C_{k+1}} |\varphi_n|^2 dx$, so $\sup_n \|\beta_k\varphi_n\|_{G/\Gamma} < \infty$ for every k. By a diagonal procedure we find a subsequence of (φ_n) such that $(\beta_k\varphi_n)_{n\in\mathbb{N}}$ converges weakly in $L^2(G/\Gamma)$ for every k. Let C be a compact subset of G. Fix k such that $(\alpha * \varphi_n)(x) = (\alpha * (\beta_k \varphi_n))(x)$ for $x \in C$, all n. Call $\beta_k \varphi_n = \psi_n$; (ψ_n) converges weakly and for any $\psi \in L^2(G/\Gamma)$ we have

$$|\alpha * \psi(x)| \leq \int_{G/\Gamma} \int_{\Gamma} |\alpha(x\gamma y^{-1})| \, |\psi(y)| dy \leq c\|x\|^r \|\psi\|_1 \leq c' \|\psi\|_2$$

for $x \in C$. Hence we conclude that $\alpha * \psi_n$ converges uniformly on C.

§3. Consequences of Theorem 6.

The notations are as in §1.

Lemma 60. Let (h_n) and (z_n) be sequences in $\underline{\underline{\Phi}}$ and D respectively such that $h_n \in \underline{\underline{\Phi}}(z_n)$. Suppose that h_n tends to h_o in $\underline{\underline{\Phi}}$ and that z_n tends to z_o in D. Then $h_o \in \underline{\underline{\Phi}}(z_o)$ and $f(h_n : z_n)$ tends to $f(h_o : z_o)$ uniformly on every compact subset of G.

Proof. By enlarging $\underline{\underline{\Phi}}$ we may assume that we have the representations $\pi(z)$ of $I_c^\infty(G)$ on $\underline{\underline{\Phi}}$ (see §2). Choose $\alpha \in I_c^\infty(G)$ such that $\pi(z_o : \alpha) = 1$. Put $h'_n = \pi(z_n : \alpha)^{-1} h_n$, $f'_n = f(h'_n : z_n)$, $f_n = \alpha * f'_n = f(h_n : z_n)$. Since f'_n remains bounded on every compact subset of G (Theorem 6), we can select a subsequence f'_{k_n} such that $\alpha * f'_{k_n} = f_{k_n}$ converges uniformly on every compact set (Lemma 59). Put $f(x) = \lim_{n\to\infty} f_{k_n}(x)$. From Theorem 6 and Lemma 54 one sees that $|f_n(x\xi)| \leq c_o e^{-\lambda(H_o(x))}$ for $x \in \mathfrak{Y}$, $\xi \in \Xi$, $n \geq 1$, hence $|f_n(x)| \leq c\|x\|^r$ for some c, r and all $x \in G$, all n ; moreover, if P is any cuspidal subgroup of G, then $|f_{n,P}(x)| \leq c_P \|x\|^r$ for $x \in G$, $n \geq 1$. From this one deduces that

$$\int_{M/\Gamma_M} (f_P(ma), \varphi(m)) dm = \lim_{n\to\infty} \int_{M/\Gamma_M} (f_{k_n,P}(ma), \varphi(m)) dm$$

for any cuspidal group $P = MAU$ and $\varphi \in {}^oL^2(M/\Gamma_M, \sigma_M, \chi)$, $\chi \in \mathcal{X}_M$. So $f \in \mathcal{A}_1(G/\Gamma, \sigma)$. On the other hand, we have

$$f^{P_i} = \lim f_{k_n}^{P_i} = \lim (h_{k_n,i})_{\chi_i}(z_{k_n}) = (h_{o,i})_{\chi_i}(z_o) \quad (1 \leq i \leq r) ;$$

so $h_o \in \underline{\underline{\Phi}}(z_o)$ and $f(h_o : z_o) = f$.

The same argument can be applied to any subsequence of the given sequence. This gives the following result: any subsequence of (f_n) contains a subsequence which converges pointwise to $f(h_o:z_o)$. Since for fixed x the sequence $(f_n(x))$ is bounded, we have therefore $\lim f_n = f(h_o:z_o)$ pointwise. Also f'_n tends pointwise to $f(h_o:z_o)$, and f'_n remains bounded on every compact subset of G, hence $\alpha * f'_n$ converges uniformly on every compact subset of G.

Corollary 1. Let $z \longmapsto h(z)$ be a continuous mapping of D into $\underline{\underline{\Phi}}$ such that $h(z) \in \underline{\underline{\Phi}}(z)$. Put $f(z) = f(h(z):z)$. Then $f(z:x)$ is continuous on $D \times G$. Moreover, given a compact subset ω of D, there exist real numbers r and c such that

$$|f(z:x)| \leq c\|x\|^r \qquad \text{for } x \in G,\ z \in \omega.$$

Proof. The first assertion follows from Lemma 60, the second one from Theorem 6 and Lemma 54.

Corollary 2. Assume that D is an open subset of $\mathbb{C}^n$ and that $\chi_i(z:a)$ is holomorphic in z $(z \in D)$ for any $a \in A_i$, $1 \leq i \leq r$. Let $z \longmapsto h(z)$ be a holomorphic mapping of D into $\underline{\underline{\Phi}}$ such that $h(z) \in \underline{\underline{\Phi}}(z)$. Define $f(z) = f(h(z):z)$. Then for any $x \in G$, $f(z:x)$ is holomorphic in z $(z \in D)$.

Proof. By Corollary 1, $f(z:x)$ is continuous on $D \times G$. Define

$$F(z:x) = \frac{1}{(2\pi i)^n} \int_{C_1} \cdots \int_{C_n} \frac{f(\zeta:x)}{(\zeta_1 - z_1)\cdots(\zeta_n - z_n)}\, d\zeta_1 \cdots d\zeta_n$$

for $x \in G$, $z = (z_j) \in D$; C_j is a (small) circle around z_j. One proves easily, using Corollary 1, that $F(z) \in \mathcal{A}_1(G/\Gamma,\sigma)$ and that $F(z)^{P_i} = (h(z)_i)_{\chi_i(z)}$. It follows then that $F(z) = f(h(z):z) = f(z)$. So $f(z:x)$ is holomorphic in z for fixed x.

§4. Langland's lemma for arbitrary q.

Fix a cuspidal subgroup $P = MAU$. Let Φ be a finite-dimensional $\mathfrak{Z}_M$-stable subspace of ${}^{o}\mathcal{A}(M/\Gamma_M,\sigma_M)$. For $N \geq 0$ we denote by $\Phi(N)$ the space spanned by all functions h on G of the form

$$h(x) = \varphi(x)p(H(x)),$$

where $\varphi \in \Phi$ and p is a polynomial function on $\mathfrak{a}_c$ with degree $\leq N$. For any $\Lambda \in \mathfrak{a}_c^*$ and $h \in \Phi(N)$, put $h_\Lambda(x) = h(x)e^{(\Lambda-\varrho)(H(x))}$ $(x \in G)$.

Put $\mathfrak{Z}_P = \mathfrak{Z}_M\mathfrak{a} + \mathfrak{G}\mathfrak{n}$; $\mathfrak{Z}_P$ is independent of A. We define for $h \in \Phi(N)$, $x \in G$:

$$\nu_P(h:x) = \int_{M/\Gamma_M} |h(xm)|\,dm,$$

$$\nu_P(\zeta:h:x) = \nu_P(\zeta h:x) \qquad (\zeta \in \mathfrak{Z}_P),$$

$$\nu_P(S:h:x) = \sum_{\zeta\in S} \nu_P(\zeta:h:x) \qquad (S \subset \mathfrak{Z}_P,\ S \text{ finite}),$$

$$\nu_P(h:\Lambda:x) = \int_{M/\Gamma_M} |h_\Lambda(xm)|\,dm = e^{\mathrm{Re}(\Lambda-\varrho)(H(x))}\nu_P(h:x) \qquad (\Lambda \in \mathfrak{a}_c^*),$$

$$\nu_P(\zeta:h:\Lambda:x) = \nu_P(\zeta h:\Lambda:x) = e^{\mathrm{Re}(\Lambda-\varrho)(H(x))}\nu_P(\zeta:h:x),$$

$$\nu_P(S:h:\Lambda:x) = \sum_{\zeta\in S} \nu_P(\zeta:h:\Lambda:x).$$

Note that $\nu_P(h:x) = \nu_P(h:a(x))$.

Let $P_1,\dots,P_r$ be a complete set of representatives for the Γ-conjugacy classes of cuspidal subgroups of G of rank q. Fix for each i a split component A_i of P_i. Let I be a finite set. For each j $(1 \leq j \leq r)$ and $i \in I$, let Φ_{ji} be a finite-dimensional $\mathfrak{Z}_{M_j}$-stable subspace of ${}^0\mathcal{A}(M_j/\Gamma_j,\sigma_j)$. Fix $N \geq 0$ and put

$$\underline{\Phi}(N) = \prod_{1\leq j\leq r}\ \prod_{i\in I} \Phi_{ji}(N), \quad \Omega = \prod_{1\leq j\leq r} (\mathfrak{a}_{j,c}^*)^I.$$

Definition. If $\Lambda = (\Lambda_{ji}) \in \Omega$, then $L(\Lambda) = \mathcal{A}_q(G/\Gamma,\sigma,\underline{\Phi}(N),\Lambda)$ is the space of all $\mathfrak{Z}$-finite functions f in $\mathcal{A}_q(G/\Gamma,\sigma)$ with the property that there exists an element $h = (h_{ji})$ of $\underline{\Phi}(N)$ such that

$$(*) \qquad f_{P_j} = \sum_{i\in I} (h_{ji})_{\Lambda_{ji}} \qquad (1 \leq j \leq r).$$

Definition. $\underline{\Phi}(N:\Lambda)$ is the space of all $h \in \underline{\Phi}(N)$ for which there exists an element f of $L(\Lambda)$ such that $(*)$ holds.

For given $h \in \underline{\Phi}(N:\Lambda)$ the element f of $L(\Lambda)$ for which $(*)$ holds is unique; write $f = f(h:\Lambda)$. The mapping $h \longmapsto f(h:\Lambda)$ is a surjective, but not necessarily bijective, map $\underline{\Phi}(N:\Lambda) \to L(\Lambda)$.

Let $\mathfrak{A}_j(N)$ be the space of all elements of $\mathfrak{A}_j$ of degree $\leq N$. Fix a base S_j^o for $\mathfrak{A}_j(N)$ $(1 \leq j \leq r)$.

Theorem 6'. Let (P_o,A_o) be a mincuspidal pair. Fix $\xi_j \in G_{\mathbb{Q}}$ such that ${}^{\xi_j}(P_j,A_j) \succ (P_o,A_o)$ $(1 \leq j \leq r)$. Then for any Siegel domain $\mathfrak{S}$ in G with respect to (P_o,A_o) and any compact set ω in Ω there exists a number c such that

$$|f(h:\Lambda:x)| \leq c \sum_{1 \leq j \leq r} \sum_{i \in I} \nu_{P_j}(S_j^o:h_{ji}:\Lambda_{ji}:x\xi_j)$$

for all $\Lambda \in \omega$, $h \in \underline{\Phi}(N:\Lambda)$, $x \in \mathfrak{S}$.

The proof of Theorem 6' is by induction on q . For $q = o$ the theorem is trivial. We shall not give the proof for $q \geq 1$ here. The proof is essentially the same as that of Theorem 6, with two differences: firstly, if $q > 1$, then the induction part is more complicated, and secondly, instead of the trivial Lemma 54 one has to prove the following

Lemma. If ω is a compact subset of Ω , then we can choose numbers c_1, c_2 with $o < c_1 \leq c_2$ and $\lambda \in \mathfrak{a}_o^*$ such that

$$c_1\mu(h)e^{\lambda(H_o(x))} \leq \sum_{j,i} \nu_{P_j}(S_j^o:h_{ji}:\Lambda_{ji}:x\xi_j) \leq c_2\mu(h)e^{-\lambda(H_o(x))}$$

for $x \in \mathfrak{S}$, $h \in \underline{\Phi}(N)$, $\Lambda \in \omega$.

(Here μ is a norm on the finite-dimensional vector space $\underline{\Phi}(N)$).

The following consequences of Theorem 6' are the analogues of Lemma 60 and its corollaries.

Corollary 1. Let (h_n) and (Λ_n) be sequences in $\underline{\Phi}(N)$ and Ω respectively such that $h_n \in \underline{\Phi}(N:\Lambda_n)$. Suppose that h_n tends to h_o in $\underline{\Phi}(N)$ and that Λ_n tends to Λ_o in Ω . Then $h_o \in \underline{\Phi}(N:\Lambda_o)$ and $f(h_n:\Lambda_n)$ tends to $f(h_o:\Lambda_o)$ uniformly on every compact subset of G .

<u>Corollary 2</u>. Let D be a topological space and $z \longmapsto h(z)$, resp. $z \longmapsto \Lambda(z)$, a continuous mapping of D into $\underline{\underline{\Phi}}(N)$, resp. Ω, such that $h(z) \in \underline{\underline{\Phi}}(N:\Lambda(z))$ for all $z \in D$. Put $f(z) = f(h(z):\Lambda(z))$. Then $f(z:x)$ is continuous on $D \times G$. Moreover, given a compact subset ω of D, there are real numbers r and c such that $|f(z:x)| \leq c\|x\|^r$ for $x \in G$, $z \in \omega$.

<u>Corollary 3</u>. In the notation of Corollary 2, assume that D is an open subset of C^n and that the mappings h and Λ are holomorphic on D. Then for any $x \in G$, $f(z:x)$ is holomorphic in z $(z \in D)$.

CHAPTER IV

§1. A partition of G/Γ.

We need a few lemmas concerning subsets of Siegel domains, which we do not prove here.

For any Siegel domain $\mathfrak{S}$ with respect to a mincuspidal pair (P,A) we introduce the following notations. Let F be a subset of $\Sigma^o(P|A)$ and t and c real numbers, $c \geq 0$.

$\mathfrak{S}(F,t) = \{x \in \mathfrak{S} : \alpha(H(x)) \leq t \quad \text{for} \quad \alpha \in {}^cF\}$,

$\mathfrak{S}(c,F) = \{x \in \mathfrak{S} : \beta(H(x)) \geq c\,\alpha(H(x)) \quad \text{for} \quad \beta \in F,\ \alpha \in {}^cF\}$,

$\mathfrak{S}(c,F,t) = \mathfrak{S}(c,F) \cap \mathfrak{S}(F,t)$.

Lemma 61. Let $\mathfrak{S}, \mathfrak{S}'$ be two Siegel domains with respect to mincuspidal pairs (P,A), (P',A') respectively. Let B, B' be compact subsets of A, A', and $F \subset \Sigma^o(P|A)$, $F' \subset \Sigma^o(P'|A')$. Fix a number c with $0 \leq c < 1$. Suppose F and F' have the same number of elements. Then there exists a real number t with the following property. If $\gamma \in \Gamma$ and $\mathfrak{S}(c,F,t)B \cap \mathfrak{S}'(c,F',t)B'\,{}^oP'\gamma$ is not empty, then ${}^{\gamma}(P_F) = P'_{F'}$.

Let $P_1,\ldots,P_r$ be a complete set of representatives for the Γ-conjugacy classes of cuspidal subgroups of G of rank 1. Fix split components A_i, so that we have the Langlands decompositions $P_i = M_iA_iU_i$ $(1 \leq i \leq r)$. For each i fix a complete set ${}^*P_{ij}$ $(1 \leq j \leq r_i)$ of representatives for the Γ_{M_i}-conjugacy classes of mincuspidal subgroups of M_i. They have rank $l-1$, if $l = \mathrm{rank}_{\mathbb{Q}}(G)$. Choose Langlands decomposition ${}^*P_{ij} = {}^*M_{ij}{}^*A_{ij}{}^*U_{ij}$. Let P_{ij} be the cuspidal subgroup of G contained in P_i which corresponds to ${}^*P_{ij}$. So $P_{ij} = M_{ij}A_{ij}U_{ij}$, $M_{ij} = {}^*M_{ij}$, $A_{ij} = {}^*A_{ij}A_i$, $U_{ij} = {}^*U_{ij}U_i$ and (P_{ij},A_{ij}) is a mincuspidal pair of G. Since (P_i,A_i) dominates (P_{ij},A_{ij}), there is a unique subset F_{ij} of $\Sigma^o(P_{ij}|A_{ij})$ such that $(P_i,A_i) = (P_{ij},A_{ij})_{F_{ij}}$.

Lemma 62. We can choose Siegel domains $\mathfrak{S}_{ij}$ with respect to (P_{ij},A_{ij}) in G

and finite subsets B^o_{ij} of A_{ij} such that $G = \bigcup_{1 \le i \le r} \bigcup_{1 \le j \le r_i} \mathfrak{F}_{ij}(1,F_{ij})B^o_{ij}\Gamma$.

Choose Siegel domains $\mathfrak{F}_{ij}$ with the property stated in Lemma 62. Put $^*\mathfrak{F}_{ij} = M_i \cap \mathfrak{F}_{ij}A_iU_i$; $^*\mathfrak{F}_{ij}$ is a Siegel domain in M_i with respect to $(^*P_{ij},{}^*A_{ij})$. We may obviously assume (after enlarging the $\mathfrak{F}_{ij}$ and the B^o_{ij}) that:

1) $1 \in B^o_{ij}$;

2) $M_i = \bigcup_{1 \le j \le r_i} {}^*\mathfrak{F}_{ij}\Gamma_{M_i}$ $(1 \le i \le r)$;

3) $\mathfrak{F}_{ij}bU_i = \mathfrak{F}_{ij}b\ (\Gamma \cap U_i)$ and $\mathfrak{F}_{ij}bU_{ij} = \mathfrak{F}_{ij}b\ (\Gamma \cap U_{ij})$ for $b \in B^o_{ij}$, $1 \le j \le r_i$, $1 \le i \le r$.

For any cuspidal pair (P,A) of rank 1 we define: $A(t) = \{a \in A : \alpha(\log a) \le t\}$ where α is the simple root of (P,A).
In particular $A_i(t)$ is defined. Put $G_i(t) = K\,A_i(t)M_iU_i$, $G_i'(t) = G_i(t)/\Gamma \cap P_i$. Fix a real number c_o with $0 \le c_o < 1$. For any real t we define

$$\tilde{\mathfrak{F}}_{ij}(t) = G_i(t) \cap \mathfrak{F}_{ij}(c_o,F_{ij})B^o_{ij} .$$

<u>Lemma 63</u>. There exist numbers τ, T and c' with $\tau \ge 0$, $T \ge 0$, $0 < c' \le 1$ such that
$$\mathfrak{F}_{ij}(c_o,F_{ij},t-T)B^o_{ij} \subset \tilde{\mathfrak{F}}_{ij}(t) \subset \mathfrak{F}_{ij}(c_o,F_{ij},c'(t+T))B^o_{ij}$$
for $t < -\tau$, $1 \le i \le r$, $1 \le j \le r_i$.

Let $\pi : G \to G/\Gamma$, $\pi_i' : G \to G/\Gamma \cap P_i$ and $\pi_i : G/\Gamma \cap P_i \to G/\Gamma$ be the canonical mappings and call $G/\Gamma \cap P_i = G_i'$. We define

$$\tilde{S}_{ij}(t) = \pi(\tilde{\mathfrak{F}}_{ij}(t)) , \quad \tilde{S}_i(t) = \bigcup_{1 \le j \le r_i} \tilde{S}_{ij}(t), \quad \tilde{S}(t) = \bigcup_{1 \le i \le r} \tilde{S}_i(t) ,$$

$$\tilde{S}'_{ij}(t) = \pi_i'(\tilde{\mathfrak{F}}_{ij}(t)), \quad \tilde{S}_i'(t) = \bigcup_{1 \le j \le r_i} \tilde{S}'_{ij}(t) .$$

<u>Lemma 64</u>. There is a number t_1 such that for all t with $t \le t_1$ we have the following. If $i \neq k$, then $\tilde{S}_i(t) \cap \tilde{S}_k(t)$ is empty; π_i defines a bijection of $\tilde{S}_i'(t)$ on $\tilde{S}_i(t)$. – This is an immediate consequence of Lemmas 61 and 63.

§2. The Maass-Selberg relations.

We first prove a few lemmas.

Lemma 65. Let Γ_o be a subgroup of Γ and $\pi_o : G \to G/\Gamma_o$ the canonical projection. Let $\mathfrak{S}$ be a Siegel domain with respect to some mincuspidal pair. If (x_n) is a sequence of elements of $\mathfrak{S}$ such that $\pi_o(x_n)$ converges in G/Γ_o, then we can select a subsequence of (x_n) which converges in G.

Proof. It is clearly enough to prove the lemma for $\Gamma_o = \Gamma$. Choose $x \in G$ such that $\lim \pi(x_n) = \pi(x)$ and choose another Siegel domain, say $\mathfrak{S}'$, with respect to the same mincuspidal pair such that $\mathfrak{S} \subset \mathfrak{S}'$ and that x lies in the interior of $\mathfrak{S}'$. There is a sequence (γ_n) of elements of Γ such that $\lim x_n\gamma_n = x$. For almost all n we have $x_n\gamma_n \in \mathfrak{S}' \cap \mathfrak{S}'\gamma_n$. Since there are but finitely many $\gamma \in \Gamma$ such that $\mathfrak{S}' \cap \mathfrak{S}'\gamma$ is not empty, we can select a subsequence such that γ_n is constant.

Corollary 1. Let Ω be a closed subset of G contained in $\mathfrak{S}$. Then $\pi_o(\Omega)$ is closed in G/Γ_o.

Corollary 2. If $\mathfrak{S}$ is closed and C is a compact subset of G/Γ_o, then $\pi_o^{-1}(C) \cap \mathfrak{S}$ is compact.

We use now the notations of §1. Whenever we speak of the sets $\tilde{S}_i(t)$ etc., it is understood that t is so small that the assertions of Lemma 64 are true.

Lemma 66. Let C be a compact subset of G/Γ. Suppose $\mathfrak{S}_{ij}$ closed (all i,j).

a) $\pi_i^{-1}(C) \cap \tilde{S}_i'(t)$ is compact.

b) $C_{ij} = \pi^{-1}(C) \cap \tilde{\mathfrak{S}}_{ij}(t)$ is compact.

c) If $C_i = \bigcup_{1 \le j \le r_i} C_{ij}$, then $\pi_i'(C_i U_i)$ is compact.

Proof. a) follows from the last assertion of Lemma 64 and Corollary 1 of Lemma 65; b) follows from Corollary 2 of Lemma 65 and c) from b).

Corollary. If $f \in C_c(G/\Gamma)$, then $\mathrm{Supp}(f_{P_i}) \cap \tilde{S}_i'(t)$ is compact.

Proof. Let C be the support of f in G/Γ and define C_{ij} and C_i as in Lemma 66. If $x \in \tilde{\mathfrak{F}}_{ij}(t)$ and $x \in \mathrm{Supp}(f_{P_i})$, then $\pi(xu) \in C$ for some $u \in U_i$, so $xu \in \tilde{\mathfrak{F}}_{ij}(t)U_i = \tilde{\mathfrak{F}}_{ij}(t)\ (\Gamma \cap U_i)$ (from the assumptions on $\mathfrak{F}_{ij}$). Write $xu = y\gamma$ with $y \in \tilde{\mathfrak{F}}_{ij}(t)$, $\gamma \in \Gamma \cap U_i$. Now $y \in C_{ij}$, so $x \in C_iU_i$ and $\pi_i^!(x) \in \pi_i^!(C_iU_i)$. This proves that $\mathrm{Supp}(f_{P_i}) \cap \tilde{S}_i^!(t) \subset \pi_i^!(C_iU_i)$ and Lemma 66c) gives the conclusion.

Lemma 67. Suppose $x \in G$ and $\pi_i^!(x) \in \tilde{S}_i^!(t)$. Then $\pi_i^!(xu) \in \tilde{S}_i^!(t)$ for any $u \in U_i$.

Proof. $x \in \tilde{\mathfrak{F}}_{ij}(t)\ (\Gamma \cap P_i)$ for some j, hence

$$xu \in \tilde{\mathfrak{F}}_{ij}(t)(\Gamma \cap P_i)U_i = \tilde{\mathfrak{F}}_{ij}(t)U_i(\Gamma \cap P_i) = \tilde{\mathfrak{F}}_{ij}(t)(\Gamma \cap P_i).$$

Let ω be the Casimir operator of $[\mathfrak{g},\mathfrak{g}]$.

Definition. $[f,g] = (\omega f,g) - (f,\omega g)$ for $f,g \in C^\infty(G,\sigma)$.

If f or g has compact support, then $\int_G [f,g]dx = 0$.

Lemma 68. Suppose $f \in C_c^\infty(G/\Gamma,\sigma)$, $g \in \mathcal{A}_1(G/\Gamma,\sigma)$ and g is $\mathfrak{Z}$-finite. Then

$$\int_{^c\tilde{S}(t)} [f,g]dx + \sum_{1\leq i\leq r} \int_{\tilde{S}_i^!(t)} [f_i^o,g_i^o]dx =$$

$$- \sum_{1\leq i\leq r} \int_{G_i^!(t)} [f_{P_i},g_{P_i}]dx + \sum_{1\leq i\leq r} \int_{G_i^!(t)\cap{}^c\tilde{S}_i^!(t)} [f_{P_i},g_{P_i}]dx$$

where $f_i^o = f - f_{P_i}$, $g_i^o = g - g_{P_i}$.

Proof. Since f has compact support, we have

$$0 = \int_{G/\Gamma} [f,g]dx = \int_{^c\tilde{S}(t)} [f,g]dx + \sum_{1\leq i\leq r} \int_{\tilde{S}_i^!(t)} [f,g]dx .$$

By the corollary of Lemma 66, $\mathrm{Supp}(f_{P_i}) \cap \tilde{S}_i^!(t)$ is compact, so we may write

$$\int_{\tilde{S}_i^!(t)} [f,g]dx = \int_{\tilde{S}_i^!(t)} [f_{P_i},g]dx + \int_{\tilde{S}_i^!(t)} [f_i^o,g]dx .$$

1) Let β_i be the characteristic function of $\tilde{S}_i'(t)$ on G_i' ; $\beta_i \circ \pi_i'$ is right-invariant under U_i (Lemma 67).

$$\int_{\tilde{S}_i'(t)} [f_{P_i}, g]dx = \int_{G_i'} \beta_i [f_{P_i}, g]dx = \int_{A_i} e^{2\varrho_i(\log a)} da \int_{M_i/\Gamma_{M_i}} \beta_i [f_{P_i}, g_{P_i}]dm =$$

$$\int_{G_i'} \beta_i [f_{P_i}, g_{P_i}]dx = \int_{\tilde{S}_i'(t)} [f_{P_i}, g_{P_i}]dx .$$

2) $$\int_{\tilde{S}_i'(t)} [f_i^o, g]dx = \int_{\tilde{S}_i'(t)} [f_i^o, g_i^o]dx + \int_{\tilde{S}_i'(t)} [f_i^o, g_{P_i}]dx$$

and in the same way as under 1) one finds

$$\int_{\tilde{S}_i'(t)} [f_i^o, g_{P_i}]dx = \int_{\tilde{S}_i'(t)} [(f_i^o)_{P_i}, g_{P_i}]dx = 0 .$$

3) $\int_{G_i'(t)} [f_{P_i}, g_{P_i}]dx$ is defined. To show this it is sufficient to prove that for $f \in C_c^\infty(G/\Gamma, \sigma)$, $g \in \mathcal{A}_1(G/\Gamma, \sigma)$, g $\mathfrak{z}$-finite, we have

$$\int_{G_i'(t)} |f_{P_i}| |g_{P_i}| \, dx < \infty .$$

Let us drop for a moment all indices i . If δ is the characteristic function of $G(t)$ on G , then

$$\int_{G'(t)} |f_P| |g_P| dx = \int_{G/\Gamma\cap P} \delta |f_P| |g_P| dx \leq \int_{G/\Gamma\cap P} \delta |f| |g_P| dx =$$

$$\int_{G/\Gamma} |f(x)| dx \sum_{\Gamma/\Gamma\cap P} \delta(x\gamma) |g_P(x\gamma)| .$$

In the last integral we have to integrate over a compact set only and one sees easily that the sum in that integral is majorized by a function E_h^* (see Lemma 26) with $h \in {}^o\mathcal{D}_{P|A}(\sigma, \xi)$ for some $\xi \subset \mathcal{X}_M$. Hence the integral is finite.

After these remarks the formula is obvious.

Now we introduce some more notations. Fix i and drop the index i ; so $(P,A) = (P_i, A_i)$, $G' = G_i' = G/\Gamma \cap P$. On G we have the Haar measure corresponding to the measure $e^{2\varrho(\log a)}\,dk\,da\,dm\,du$ on $K \times A \times M \times U$, where dk and du are such that K and $U/U \cap \Gamma$ have volume 1 and where da corresponds to the length in $\mathfrak{a}$. Choose $H \in \mathfrak{a}$ such that $\alpha(H) > 0$, $|H| = 1$, α being the simple root of (P,A). Define

$$a_t = \exp tH\,, \quad f_P(t:m) = e^{t\varrho(H)} f_P(a_t m) \qquad (t \in R,\ m \in M)\,.$$

We are going to compute $\int_{G'(T)} [f_P, g_P]dx$ (f and g as in Lemma 68).

Let μ be the canonical homomorphism of $\mathfrak{Z}$ into $\mathfrak{Z}_M \mathfrak{A}$. When one extends $\mathfrak{a}$ to a Cartan subalgebra of $\mathfrak{g}$, one sees that

$$\mu(\omega) = \omega_{\mathfrak{m}} + H^2 - \varrho(H)^2$$

where $\omega_{\mathfrak{m}}$ is a self-adjoint element of $\mathfrak{Z}_M$. We have

$$\int_{G'(T)} [f_P, g_P]dx = \int_{-\infty}^{T/|\alpha|} e^{2t\varrho(H)}dt \int_{M/\Gamma_M} F(a_t m)\,dm\,,$$

where $F = [f_P, g_P] = (\mu(\omega)' f_P, g_P) - (f_P, \mu(\omega)' g_P)$. Put $u_o = H^2 - \varrho(H)^2$; then $\mu(\omega)' = \omega_{\mathfrak{m}} + u_o'$. We have

$$\int_{M/\Gamma_M} \{(\omega_{\mathfrak{m}} f_P, g_P) - (f_P, \omega_{\mathfrak{m}} g_P)\}dm = o$$

and, since $(u_o' f_P)(a_t m) = e^{-t\varrho(H)}\left(\frac{d^2}{dt^2} - \varrho(H)^2\right) f_P(t:m)$,

$$\int_{M/\Gamma_M} \{(u_o' f_P, g_P) - (f_P, u_o' g_P)\}dm =$$

$$e^{-2t\varrho(H)} \int_{M/\Gamma_M} \{(\frac{d^2}{dt^2} f_P(t:m), g_P(t:m)) - (f_P(t:m), \frac{d^2}{dt^2} g_P(t:m))\}dm\,.$$

So we obtain

$$\int_{G'(T)} [f_P, g_P]dx = \int_{-\infty}^{T/|\alpha|} \frac{d}{dt} J_P(f,g:t)\,dt\,,$$

where

$$J_P(f,g:t) = \int_{M/\Gamma_M} \{(\frac{d}{dt} f_P(t:m), g_P(t:m)) - (f_P(t:m), \frac{d}{dt} g_P(t:m))\}dm\,.$$

From the assumption that f has compact support modulo Γ one deduces that $\alpha(H_{P|A}(x))$ is bounded from below on $\mathrm{Supp}(f_P)$; hence $f_P(t:m) = o$ and $J_P(f,g:t) = o$ for t sufficiently large and negative. So

$$\int_{G'(T)} [f_P,g_P]dx = J_P(f,g:T/|\alpha|) .$$

Putting this result into the formula of Lemma 68 we get:

<u>Lemma 69</u>. For $f \in C_c^m(G/\Gamma,\sigma)$, $g \in \mathcal{A}_1(G/\Gamma,\sigma)$, g $\mathfrak{Z}$-finite, we have

$$\int_{{}^c\tilde{S}(t)} [f,g]dx + \sum_{1\leq i\leq r} \int_{\tilde{S}_i'(t)} [f_i^o,g_i^o]dx =$$

$$- \sum_{1\leq i\leq r} J_{P_i}(f,g:\frac{t}{|\alpha_i|}) + \sum_{1\leq i\leq r} \int_{G_i'(t)\cap{}^c\tilde{S}_i'(t)} [f_{P_i},g_{P_i}]dx .$$

The same formula is valid for two $\mathfrak{Z}$-finite functions f, g in $\mathcal{A}_1(G/\Gamma,\sigma)$. This is the <u>Maass-Selberg relation</u> (see Maass, Math. Ann. vol. 121 (1949), pp. 141-183). We shall not prove it in extenso here. It can be derived from Lemma 69 by a limit process: Let f,g be two $\mathfrak{Z}$-finite functions in $\mathcal{A}_1(G/\Gamma,\sigma)$. Fix $\alpha \in I_c^\infty(G)$ such that $\alpha * f = f$. Let $\Omega_o = \Omega_o^{-1}$ be a compact meighbourhood of 1 in G such that $\mathrm{Supp}(\alpha) \subset \Omega_o$. Choose compact sets $\omega_i \subset U_i$ such that $\omega_i(U_i\cap\Gamma_i) = U_i$. Let C_k $(k \geq o)$ be a sequence of compact sets in G such that 1) $\Omega_o C_k \omega_i \subset C_{k+1}$, 2) $\cup\, C_k = G$. Let β_k be the characteristic function of $\pi(C_k)$ on G/Γ . Put $f_k = \alpha*(\beta_k f)$. Apply Lemma 69 to f_k,g and let k tend to infinity.

In the described procedure the following lemma, which we state only, is to be used for the estimation of integrals.

Let (P,A) be a cuspidal pair of rank 1 and (P_o,A_o) a mincuspidal pair dominated by (P,A). Let α be the unique root in $\Sigma^o(P_o|A_o)$ which does not vanish identically on A . Let $\mathfrak{F}$ be a Siegel domain with respect to (P_o,A_o) . For any function f on G and any $\lambda \in \mathfrak{a}_o^*$ define

$$\nu_\lambda(A:f) = \sup_{y \in \mathfrak{F}AU_o} |f(y)| e^{\lambda(H_o(y))} .$$

For any $\mu \in \mathfrak{a}^*$ let $\mathfrak{a}_o^*(\mu)$ denote the set of all $\lambda \in \mathfrak{a}_o^*$ such that $\lambda|_{\mathfrak{a}} = \mu$. Put ${}^*\mathfrak{F} = M \cap \mathfrak{F}AU$. Fix $c_o > o$ and put

$${}^*\mathfrak{F}(a) = \{m \in {}^*\mathfrak{F} : ma \notin \mathfrak{F}(c_o, {}^c\{a\})U_o\} .$$

Lemma. Fix $\mu \in \mathfrak{a}^*$. Given $\lambda_o \in \mathfrak{a}_o^*$, $r_o \geq o$, there exist $\lambda \in \mathfrak{a}_o^*(\mu)$, $r \geq r_o$, $c > o$, such that

$$|f(am)| e^{\lambda_o(H_o(am))} \leq c\, \nu_\lambda(A:f) e^{r\alpha(\log a)}$$

for $a \in A$, $m \in {}^*\mathfrak{F}(a)$ and any function f on G.

§3. The functions $\check{E}_h$.

Let $P = MAU$ be any cuspidal subgroup of G and Φ some space ${}^oL^2(M/\Gamma_M, \sigma_M, \xi)$ where ξ is a finite subset of $\mathcal{H}_M$ which we suppose stable under $w(\mathfrak{a})$. If v is an element of $C_c^\infty(A) \otimes \Phi$, we consider it as a function on A with values in Φ and put $f_v(x) = (v(a(x)))(x) = v(a(x):x)$. Then $f_v \in {}^o\mathcal{D}_{P|A}(\sigma,\xi)$ (Chap. II,§6) and the mapping $v \longmapsto f_v$ is a bijection of $C_c^\infty(A) \otimes \Phi$ on ${}^o\mathcal{D}_{P|A}(\sigma,\xi)$. If, in particular $v = w \otimes \varphi$ with $w \in C_c^\infty(A)$, $\varphi \in \Phi$, then $f_v(x) = w(a(x))\varphi(x)$. For any $v \in C_c^\infty(A) \otimes \Phi$ we put $E_v = E_{f_v}$. The Fourier transform of v is defined by

$$\hat{v}(\Lambda) = \int_A v(a) e^{-(\Lambda-\varrho)(\log a)} da .$$

We have $\hat{v}(\Lambda:x) = \hat{f}_v(\Lambda:x)$ if $\hat{f}_v$ is defined as in Chap. II,§3.

For any $\lambda \in \mathfrak{a}^*$ define

$$q_\lambda(a) = e^{2\varrho(\log a)} \sum_{s \in w(\mathfrak{a})} (e^{2s\lambda(\log a)} + e^{-2s\lambda(\log a)}) \quad (a \in A).$$

and denote by $L^2(A:\lambda)$ the space of all functions on A which are square integrable with respect to the measure $q_\lambda(a)da$. Put

$$\nu_\lambda(v)^2 = \int_A |v(a)|^2 \, q_\lambda(a) da$$

for any measurable function $v : A \to \Phi$.

Lemma 70. Let ω^* be a compact subset of $(\mathfrak{a}^*)^-$. There is a number c such that

$$\|E_v\|_{G/\Gamma} \leq c\, \nu_\lambda(v)$$

for all $\lambda \in \omega^*$, $v \in C_c^\infty(A) \otimes \Phi$.

Proof. By Lemma 40 we have for any $\lambda \in (\mathfrak{a}^*)^-$

$$\|E_v\|^2_{G/\Gamma} = \sum_{s \in W(\mathfrak{a})} \int_{\mathfrak{a}^*} (\hat{v}(-s\bar{\Lambda}), c(s:\Lambda)\hat{v}(\Lambda))_\Phi \, d\Lambda_I \qquad (\Lambda_R = \lambda).$$

By Lemma 38, $|c(s:\Lambda)|$ is bounded for $\Lambda_R = \lambda \in \omega^*$, hence

$$\|E_v\|^2 \leq c_0 \sum (\int |\hat{v}(-s\bar{\Lambda})|^2 d\Lambda_I)^{1/2} \, (\int |\hat{v}(\Lambda)|^2 d\Lambda_I)^{1/2} .$$

$\int_{\mathfrak{a}^*} |\hat{v}(\Lambda)|^2 d\Lambda_I = \int_A |v(a)|^2 \, e^{-2(\lambda-\varrho)(\log a)} da \leq \nu_\lambda(v)^2$ and similarly for the other integral.

Lemma 70 implies that the mapping $v \longmapsto E_v$ of $C_c^\infty(A) \otimes \Phi$ into $L^2(G/\Gamma,\sigma)$ can be extended in a unique way to a continuous linear map $L^2(A:\lambda) \otimes \Phi \to L^2(G/\Gamma,\sigma)$ if $\lambda \in (\mathfrak{a}^*)^-$. We use the same notation for the extended map.

Lemma 71. If $v_1, v_2 \in L^2(A:\lambda) \otimes \Phi$ and $\lambda \in (\mathfrak{a}^*)^-$, then

$$((E_{v_2}, E_{v_1}))_{G/\Gamma} = \sum_{s \in W(\mathfrak{a})} \int_{\mathfrak{a}^*} (\hat{v}_2(-s\bar{\Lambda}), \; c(s:\Lambda)\hat{v}_1(\Lambda))_\Phi \, d\Lambda_I \qquad (\Lambda_R = \lambda).$$

Proof. This equality holds if $v_1, v_2 \in C_c^\infty(A) \otimes \Phi$ (Lemma 40) and both sides are continuous functions of v_1, v_2 (cf. the proof of Lemma 70 for the continuity of the right hand side).

Fix a real number r with $r > |\varrho|$. Let $\mathfrak{a}^*(r)$ denote the set of all λ in $\mathfrak{a}^*$ such that $|\lambda| < r$ and $\mathfrak{a}_c^*(r)$ the set of all Λ in $\mathfrak{a}_c^*$ such that $\mathrm{Re}(\Lambda) \in \mathfrak{a}^*(r)$. Obviously $\mathfrak{a}^*(r) \cap (\mathfrak{a}^*)^-$ is not empty.

<u>Definition</u>. $\mathcal{H}_r$ is the set of all bounded holomorphic functions h on $\mathfrak{a}_c^*(r)$ such that $\int_{\mathfrak{a}^*} |h(\lambda+i\mu)|^2 d\mu < \infty$ for all $\lambda \in \mathfrak{a}^*(r)$.

$\mathcal{H}_r$ contains the Fourier transforms of the functions of $C_c^\infty(A)$.

Fix $h \in \mathcal{H}_r$ and let, for $\lambda \in \mathfrak{a}^*(r)$, g_λ be the function in $L^2(A)$ which is the Fourier transform of the function $\mu \longmapsto h(\lambda+i\mu)$ on $\mathfrak{a}^*$ ($g_\lambda(a) = \int_{\mathfrak{a}^*} h(\lambda+i\mu)e^{i\mu(\log a)}d\mu$ for as far as that integral makes sense). For $v \in C_c^\infty(A)$ we have then

$$(*) \qquad \int_A v(a)e^{-(\lambda-\varrho)(\log a)} g_\lambda(a^{-1})da = \int_{\mathfrak{a}^*} \hat{v}(\lambda+i\mu)h(\lambda+i\mu)d\mu .$$

It is clear that the integral

$$\int_{\mathfrak{a}^*} |\hat{v}(\Lambda+i\mu)h(\Lambda+i\mu)|d\mu$$

converges uniformly for Λ in a compact subset of $\mathfrak{a}_c^*(r)$; hence

$$\int_{\mathfrak{a}^*} \hat{v}(\Lambda+i\mu)h(\Lambda+i\mu)d\mu$$

is a holomorphic function of Λ on $\mathfrak{a}_c^*(r)$, and since it is independent of $\mathrm{Im}(\Lambda)$, it is constant. From $(*)$ we see then that there is a function g on A such that for every $\lambda \in \mathfrak{a}^*(r)$ we have

$$g_\lambda(a)e^{(\lambda-\varrho)(\log a)} = g(a) \quad \text{almost everywhere.}$$

Since $g_\lambda \in L^2(A)$ and $|s\lambda| = |\lambda|$ for $s \in w(\mathfrak{a})$, the function g belongs to $L^2(A:\lambda)$ for any $\lambda \in \mathfrak{a}^*(r)$. Thus we have constructed, corresponding to $h \in \mathcal{H}_r$, a function g in $L^2(A:\lambda)$. If $\varphi \in \Phi$, then to $h \otimes \varphi$ we make correspond $g \otimes \varphi$, and generally, to any $h \in \mathcal{H}_r \otimes \Phi$ (which can be identified with the set of all bounded holomorphic functions on $\mathfrak{a}_c^*(r)$ with values in Φ such that $\int |h(\lambda+i\mu)|^2 d\mu < \infty$ for all $\lambda \in \mathfrak{a}^*(r)$) corresponds an almost everywhere defined function g on A with values in Φ such that $g \in L^2(A:\lambda) \otimes \Phi$ for any $\lambda \in \mathfrak{a}^*(r)$.

Suppose $h \in \mathcal{H}_r \otimes \Phi$ and let g correspond to h as above. Fix $\lambda \in \mathfrak{a}^*(r) \cap (\mathfrak{a}^*)^-$. Then $g \in L^2(A:\lambda) \otimes \Phi$ and we can define E_g (Lemma 70);

$E_g \in \mathrm{Cl}\,(E_{P|A}(\sigma,\xi))$ (notation of Chap. II,§6). We claim that E_g is independent of the choice of λ. To prove this it is sufficient to show that $((E_v, E_g))_{G/\Gamma}$ is independent of λ $(v \in C_c^\infty(A) \otimes \Phi)$. By Lemma 71 we have

$$((E_v, E_g))_{G/\Gamma} = \sum \int (\hat{v}(-s\bar{\Lambda}),\ c(s:\Lambda)h(\Lambda))_\Phi d\Lambda_I$$

where $\Lambda_R = \lambda$; these integrals are independent of λ. Now we define $\check{E}_h = E_g$. So $\check{E}_h$ is an element of $L^2(G/\Gamma,\sigma)$ which lies in the closure of $E_{P|A}(\sigma,\xi)$. The scalar-product formula for these functions reads

$$((\check{E}_{h_2}, \check{E}_{h_1}))_{G/\Gamma} = \sum_{s \in w(\mathfrak{a})} \int_{\mathfrak{a}^*} (h_2(-s\bar{\Lambda}), c(s:\Lambda)h_1(\Lambda))_\Phi \, d\Lambda_I$$

where $\Lambda_R \in \mathfrak{a}^*(r) \cap (\mathfrak{a}^*)^-$, $h_1, h_2 \in \mathcal{H}_r \otimes \Phi$.

From now on we shall <u>write $\mathcal{H}_r$ instead of</u> $\mathcal{H}_r \otimes \Phi$. Let f be a bounded holomorphic function $\mathfrak{a}_c^*(r) \to \mathbb{C}$. Put $f_*(\Lambda) = \overline{f(-\bar{\Lambda})}$. If $h \in \mathcal{H}_r$, then $fh \in \mathcal{H}_r$ and if $h_1, h_2 \in \mathcal{H}_r$, then

$$((\check{E}_{h_1}, \check{E}_{fh_2}))_{G/\Gamma} = ((\check{E}_{f_* h_1}, \check{E}_{h_2}))_{G/\Gamma}$$

provided f is invariant under $w(\mathfrak{a})$. One sees from this that $\check{E}_{fh} = 0$ if $\check{E}_h = 0$.

<u>Lemma 72</u>. Let f be a bounded holomorphic function on $\mathfrak{a}_c^*(r)$ such that $f(s\Lambda) = f(\Lambda)$ for $s \in w(\mathfrak{a})$. There exists a unique bounded linear transformation T_f on $\mathrm{Cl}\,(E_{P|A}(\sigma,\xi))$ such that $T_f\check{E}_h = \check{E}_{fh}$ for $h \in \mathcal{H}_r$. Moreover $\|T_f\| \leq \|f\|_\infty = \sup_{\Lambda \in \mathfrak{a}_c^*(r)} |f(\Lambda)|$.

<u>Proof</u>. Only the boundedness and the bound of T_f require proof. Fix $N > \|f\|_\infty$ and put $N^2 - f_* f = g$. We can obviously choose a holomorphic function g_1 such that $g_1^2 = g$. It is also trivial that g_1 is invariant under $w(\mathfrak{a})$ and that $(g_1)_* = g_1$. So $\|\check{E}_{g_1 h}\|^2 = ((\check{E}_{g_1 h}, \check{E}_{g_1 h})) = ((\check{E}_h, \check{E}_{gh})) = N^2\|\check{E}_h\|^2 - ((\check{E}_h, \check{E}_{f_* fh})) = N^2\|\check{E}_h\|^2 - \|\check{E}_{fh}\|^2$. Hence $\|\check{E}_{fh}\| \leq N\|\check{E}_h\|$.

We want to find out how $\mathfrak{Z}$ acts on the functions $\check{E}_h$. The assumption that ξ is stable under $w(\mathfrak{a})$ can be dropped for the moment (until Lemma 75).

Define for $\alpha \in C_c(G)$

$$T_\alpha(x) = \int\int_{K\times U} \sigma(k^{-1})\alpha(kxu)\,dk\,du \qquad (x \in G).$$

T_α is a function with values in End(V), it satisfies $T_\alpha(kxu) = \sigma(k)T_\alpha(x)$ $(k \in K, x \in G, u \in U)$ and has compact support modulo U. Assume that ${}^k\alpha = \alpha$ for $k \in K$. Then we have for any $f \in C\,(G/U,\sigma)$

$$(\alpha * f)(m_o a_o) = \int\int_{M\times A} T_\alpha(ma) f(m^{-1}\, m_o\, a^{-1}\, a_o)\,dm\,da \qquad (m_o \in M,\ a_o \in A).$$

Fix $a \in A$ and put $\gamma(m) = T_\alpha(ma)$. Then $\gamma \in C_c(M) \otimes \mathrm{End}(V)$ and $\gamma(km) = \sigma_M(k)\gamma(m)$ $(k \in K_M,\ m \in M)$. So $\varphi \longmapsto \gamma * \varphi$ defines an endomorphism of Φ, which we denote by $\pi(\alpha:a)$. We have $\pi(\alpha) \in C_c(A) \otimes \mathrm{End}(\Phi)$. The following lemma is nothing but a new formulation of Lemma 41.

Lemma 73. Let $\alpha \in C_c(G)$ such that ${}^k\alpha = \alpha$ for $k \in K$ and let $v \in C_c^\infty(A) \otimes \Phi$. Then $\alpha * f_v = f_{\pi(\alpha)*v}$ and $\alpha * E_v = E_{\pi(\alpha)*v}$.

Let $\hat{\pi}(\alpha)$ denote the Fourier transform of $\pi(\alpha)$, i.e.

$$\hat{\pi}(\alpha:\Lambda) = \int_A \pi(\alpha:a)\, e^{-(\Lambda-\varrho)(\log a)}\, da .$$

Then $\hat{\pi}(\alpha:\Lambda)$ is what we denoted by $\pi(\chi:\alpha)$ in Chap. III, §1, if one takes there $\chi(a) = e^{(\Lambda-\varrho)(\log a)}$. After what has been said l.c. we have for $\varphi \in \Phi$,

$$\hat{\pi}(\alpha:\Lambda)\varphi = \alpha^\Lambda * \varphi ,$$

where

$$\alpha^\Lambda(m) = \int_A T_\alpha(ma)\, e^{-(\Lambda-\varrho)(\log a)}\, da$$

$(\alpha^\Lambda(m) = \tau_\alpha(m)$ in the notation of l.c.).

Lemma 74. If $\alpha \in I_c^\infty(G)$ and $z \in \mathfrak{Z}$, then $(z\alpha)^\Lambda = \mu(z:\Lambda)\alpha^\Lambda$, where μ is the canonical homomorphism $\mathfrak{Z} \to \mathfrak{Z}_M$ $\mathfrak{a}$.

Proof. $T_{z\alpha}(x) = T_\alpha(x:\mu(z)')$, etc.

Lemma 75. If $\alpha \in C_c(G)$ such that ${}^k\alpha = \alpha$ for $k \in K$, and $h \in \mathcal{H}_r$, then $\alpha * \check{E}_h = \check{E}_{\hat{\pi}(\alpha)h}$.

Proof. Let g be the function which was used to define $\check{E}_h$. Choose $\lambda \in \mathfrak{a}^*(r) \cap (\mathfrak{a}^*)^-$. Then $\check{E}_h = E_g = \lim E_v$ where $v \in C_c^\infty(A) \otimes \Phi$ and $\lim v = g$ in $L^2(A:\lambda) \otimes \Phi$. We have $\alpha * \check{E}_h = \lim \alpha * E_v$, hence for any $u \in C_c^\infty(A) \otimes \Phi$,

$$((E_u, \alpha * \check{E}_h)) = \lim((E_u, \alpha * E_v)) = \lim((E_u, E_{\pi(\alpha)*v})) .$$

The scalar-product formula gives

$$((E_u, E_{\pi(\alpha)*v})) = \sum \int (\hat{u}(-s\bar{\Lambda}), c(s:\Lambda)\hat{\pi}(\alpha:\Lambda)\hat{v}(\Lambda))_\Phi \, d\Lambda_I$$

where $\Lambda_R = \lambda$. From $\lim v = g$ in $L^2(A:\lambda) \otimes \Phi$ it follows that

$$\lim \int_{\mathfrak{a}^*} |\hat{v}(\lambda+i\mu) - h(\lambda+i\mu)|^2 \, d\mu = 0 ,$$

hence, by the scalar-product formula, $\lim((E_u, E_{\pi(\alpha)*v})) = ((E_u, \check{E}_{\hat{\pi}(\alpha)h}))$. This proves Lemma 75.

Let $\varphi \in \Phi$ and $\zeta \in \mathfrak{Z}_M$. Write $\varphi = \sum_{\chi \in \xi} \varphi_\chi$ with $\varphi_\chi \in {}^oL^2(M/\Gamma_M, \sigma_M, \chi)$. Then $\zeta\varphi = \sum_{\chi \in \xi} \chi(\zeta)\varphi_\chi = \xi(\zeta)\varphi$ where ξ denotes a representation of $\mathfrak{Z}_M$ on Φ. In particular, for $z \in \mathfrak{Z}$, $\Lambda \in \mathfrak{a}_c^*$,

$$\mu(z:\Lambda)\varphi = \xi(z:\Lambda)\varphi ,$$

where $\xi(z:\Lambda) = \xi(\mu(z:\Lambda))$.

Definition. $\mathcal{H}_r^o$ is the set of all holomorphic functions h on $\mathfrak{a}_c^*(r)$ with values in Φ such that ph is bounded for every polynomial function p on $\mathfrak{a}_c^*$.

$\mathcal{H}_r^o$ is a subspace of $\mathcal{H}_r$ which contains the Fourier transforms of the functions in $C_c^\infty(A) \otimes \Phi$. If $h \in \mathcal{H}_r^o$, then the function $\Lambda \longmapsto \xi(z:\Lambda)h(\Lambda)$ lies in $\mathcal{H}_r$.

Lemma 76. If $\alpha \in I_c^\infty(G)$, $h \in \mathcal{H}_r^o$, $z \in \mathfrak{Z}$, then $z\alpha * \check{E}_h = \alpha * \check{E}_{h_z}$, where $h_z(\Lambda) = \xi(z:\Lambda)h(\Lambda)$.

Proof. By Lemma 75, $z\alpha * \check{E}_h = \check{E}_{\hat{\pi}(z\alpha)h}$. Now $(\hat{\pi}(z\alpha)h)(\Lambda) = (z\alpha)^{\Lambda} * h(\Lambda) =$ $= (\mu(z:\Lambda)\alpha^{\Lambda}) * h(\Lambda) = \alpha^{\Lambda} * (\mu(z:\Lambda)h(\Lambda)) = \alpha^{\Lambda} * h_z(\Lambda) = (\hat{\pi}(\alpha)h_z)(\Lambda)$. Apply once more Lemma 75.

We recall a few facts about representations of Lie groups.

Let π be a unitary representation of a unimodular Lie group G on a Hilbert space $\mathcal{H}$. The subspace $\mathcal{H}_\infty$ of all differentiable elements of $\mathcal{H}$ (i.e. the space of all $\varphi \in \mathcal{H}$ such that $x \longmapsto \pi(x)\varphi$ is a C^∞ map $G \to \mathcal{H}$) is dense in $\mathcal{H}$ and we have a representation π_∞ of $\mathfrak{g}$ on $\mathcal{H}_\infty$. For $g \in \mathfrak{g}$ we denote the adjoint of g by g^*, and if η is the conjugation of $\mathfrak{g}_c$ with respect to $\mathfrak{g}$, we put $g^+ = \eta(g^*)$. Then $(\varphi, \pi_\infty(g)\psi) = (\pi_\infty(g^+)\varphi, \psi)$ for $\varphi, \psi \in \mathcal{H}_\infty$, $g \in \mathfrak{g}$. An element g of $\mathfrak{g}$ is called hermitian if $g^+ = g$. If z is a hermitian element of $\mathfrak{z}$, then $\pi_\infty(z)$ is essentially self-adjoint (i.e. admits a unique self-adjoint extension).

We apply this to the representation λ of our group G on $L^2(G/\Gamma)$. All $f \in C_c^\infty(G/\Gamma)$ are differentiable vectors with respect to this representation and for $f \in C_c^\infty(G/\Gamma)$, $g \in \mathfrak{g}$, we have $(\lambda_\infty(g)f)(x) = f(g^* ; x)$. If $z \in \mathfrak{z}$, then $\lambda_\infty(z^*)f = zf$ for $f \in C_c^\infty(G/\Gamma)$. We shall denote by z also the closure of the operator $\lambda_\infty(z^*)$. Then we have

Lemma 77. If $h \in \mathcal{H}_r^o$, $z \in \mathfrak{z}$, then $z\check{E}_h = \check{E}_{h_z}$, where h_z is defined as in Lemma 76.

Proof. Choose $\alpha \in I_c^\infty(G)$. Then $\alpha * \check{E}_h$ is a differentiable vector and $z(\alpha * \check{E}_h) =$ $= (z\alpha) * \check{E}_h = \alpha * \check{E}_{h_z}$ by Lemma 76. When we let α tend to the Dirac measure concentrated at the point 1, we get $z\check{E}_h = \check{E}_{h_z}$.

§4. Further preparation of the analytic continuation.

Let (P,A) and (P,A') be two cuspidal pairs, $P = MAU$. Then we can choose $y \in P_{\mathbb{Q}}$ such that ${}^yA = A'$. The coset $yM_{\mathbb{Q}}A_{\mathbb{Q}}$ is determined by A'. The Langlands decomposition of P corresponding to A' is $P = M'A'U$ with $M' = {}^yM$. The bijection of $\mathfrak{z}_M$ on $\mathfrak{z}_{M'}$ induced by $Ad(y)$ is independent of the choice of y in $P_{\mathbb{Q}}$; so we get a bijection of $\mathcal{H}_M$ on $\mathcal{H}_{M'}$ by means of which we identify those two sets and call them $\mathcal{H}(P)$. Moreover the Weyl groups $w(\mathfrak{a})$, $w(\mathfrak{a}')$ operate on $\mathcal{H}_M$, $\mathcal{H}_{M'}$ respectively, and y defines an isomorphism of $w(\mathfrak{a})$ on $w(\mathfrak{a}')$ which is independent of the choice of y, so we identify those groups and call them w_P. This identification is compatible with the identification of the sets $\mathcal{H}_M$ and $\mathcal{H}_{M'}$ on which $w(\mathfrak{a})$ and $w(\mathfrak{a}')$ operate. So w_P operates on $\mathcal{H}(P)$; we denote by $\mathcal{H}^o(P)$ the set of orbits of w_P in $\mathcal{H}(P)$.

If $\Phi = {}^oL^2(M/\Gamma_M,\sigma_M,\chi)$ and $\Phi' = {}^oL^2(M'/\Gamma_{M'},\sigma_{M'},\chi')$ where χ in $\mathcal{H}_M$ and χ' in $\mathcal{H}_{M'}$ correspond to each other, then after extending the functions in the usual way to G, we have $\Phi = \Phi'$. Also ${}^o\mathcal{D}_{P|A}(\sigma,\chi) = {}^o\mathcal{D}_{P|A'}(\sigma,\chi')$; we write ${}^o\mathcal{D}_P(\sigma,\chi)$, where χ has to be understood as an element of $\mathcal{H}(P)$. For $\xi \in \mathcal{H}^o(P)$ we define

$$ {}^o\mathcal{D}_P(\sigma,\xi) = \sum_{\chi\in\xi} {}^o\mathcal{D}_P(\sigma,\chi) . $$

The space of all functions E_f with $f \in {}^o\mathcal{D}_P(\sigma,\chi)$ $(\chi \in \mathcal{H}(P))$ will be denoted by $E_P(\sigma,\chi)$ and for $\xi \in \mathcal{H}^o(P)$ we put

$$ E_P(\sigma,\xi) = \sum_{\chi\in\xi} E_P(\sigma,\chi) . $$

Assume now that P and P' are two cuspidal groups which are conjugate under $G_{\mathbb{Q}}$. Choose split components A and A' in P and P'. Choose $y \in G_{\mathbb{Q}}$ such that ${}^y(P,A) = (P',A')$. The coset $yM_{\mathbb{Q}}A_{\mathbb{Q}}$ is then determined by (P',A'). Again $\mathrm{Ad}(y)$ induces an isomorphism of $\mathfrak{Z}_M$ on $\mathfrak{Z}_{M'}$ which is independent of the choice of y in $G_{\mathbb{Q}}$ and so we get an isomorphism of $\mathcal{H}_M$ on $\mathcal{H}_{M'}$. When A and A' vary, these isomorphisms are compatible with the identification of the various $\mathcal{H}_M$, resp. $\mathcal{H}_{M'}$; so we have canonical isomorphisms

$$ i_{P|P'} : \mathcal{H}(P) \to \mathcal{H}(P') $$

whenever P and P' are conjugate. Moreover, in the above notation, y defines an isomorphism between the Weyl groups $w(\mathfrak{a})$ and $w(\mathfrak{a}')$, which gives an isomorphism $w_P \to w_{P'}$ such that $i_{P|P'}$ is an isomorphism of sets with operators. Hence there is a canonical bijection

$$ \mathcal{H}^o(P) \to \mathcal{H}^o(P') $$

which we shall denote also by $i_{P|P'}$.

If P and P' are even conjugate under Γ, we have $E_P(\sigma,\chi) = E_{P'}(\sigma,\chi')$ for $\chi \in \mathcal{H}(P)$, $\chi' = i_{P|P'}(\chi)$.

Now assume that P and P' are only associated. Choose split components A and A'. Choose $y \in G_{\mathbb{Q}}$ such that ${}^yA = A'$. Now only $y\,N(A)_{\mathbb{Q}}$ is determined by A', so that the mapping $\chi \longmapsto {}^y\chi$ of $\mathcal{H}_M$ on $\mathcal{H}_{M'}$ depends on y. There is however a canonical bijection

$$i_{P|P'} : \mathcal{H}^\circ(P) \to \mathcal{H}^\circ(P') .$$

If $\xi \in \mathcal{H}^\circ(P)$, $\xi' = i_{P|P'}(\xi)$, we call ξ and ξ' associated.

If $\mathcal{C}$ is a class of associated cuspidal subgroups of G, we denote by $\mathcal{H}^\circ(\mathcal{C})$ the subset of $\prod_{P\in\mathcal{C}} \mathcal{H}^\circ(P)$ consisting of all families $\xi = (\xi_P)$ such that ξ_P and $\xi_{P'}$ are associated for every $P,P' \in \mathcal{C}$. If $\xi \in \mathcal{H}^\circ(\mathcal{C})$, we put

$$E_{\mathcal{C}}(\sigma,\xi) = \sum_{P\in\mathcal{C}} E_P(\sigma,\xi_P) = \sum_{P\in\mathcal{C}/\Gamma} E_P(\sigma,\xi_P) .$$

If $\mathcal{C}_1$ and $\mathcal{C}_2$ are two classes of associated cuspidal subgroups and $\xi_i \in \mathcal{H}^\circ(\mathcal{C}_i)$ $(i = 1,2)$, then $E_{\mathcal{C}_1}(\sigma,\xi_1)$ and $E_{\mathcal{C}_2}(\sigma,\xi_2)$ are orthogonal unless $\mathcal{C}_1 = \mathcal{C}_2$, $\xi_1 = \xi_2$ (Lemmas 39 and 40).

Fix a class $\mathcal{C}$ of associated cuspidal groups. Fix also a mincuspidal pair (P_o,A_o). Choose a complete set of representatives $P_1,\dots,P_r$ for $\mathcal{C}/G_{\mathbb{Q}}$ such that $P_k \supset P_o$ (all k). Then $P_k = (P_o)_{F_k}$ with $F_k \subset \Sigma^o(P_o|A_o)$; put $A_k = (A_o)_{F_k}$. Let $\mathcal{C}_k$ denote the class of cuspidal groups which are conjugate to P_k ; then $\mathcal{C}$ is the disjoint union of the sets $\mathcal{C}_k$. Choose for each k a complete set of representatives P_{kl} $(1 \le l \le r_k)$ for $\mathcal{C}_k/\Gamma$ and fix y_{kl} in $G_{\mathbb{Q}}$ such that $P_{kl} = {}^{y_{kl}}(P_k)$. The set of all P_{kl} $(1 \le k \le r,\ 1 \le l \le r_k)$ is then a complete set of representatives for $\mathcal{C}/\Gamma$. Put $A_{kl} = {}^{y_{kl}}(A_k)$ and let $P_k = M_kA_kU_k$, $P_{kl} = M_{kl}A_{kl}U_{kl}$ be the Langlands decompositions.

If $\xi \in \mathcal{H}^\circ(\mathcal{C})$, we introduce the following abbreviations:

$${}^o\mathcal{D}_{kl}(\sigma,\xi) = {}^o\mathcal{D}_{P_{kl}}(\sigma,\xi_{kl}), \quad \xi_{kl} = \xi_{P_{kl}} ,$$

$$L_{kl}(\sigma,\xi) = {}^{0}L^{2}(M_{kl}/\Gamma_{kl},\sigma_{kl},\xi_{kl}),\ \Gamma_{kl} = \Gamma_{M_{kl}},\sigma_{kl} = \sigma_{M_{kl}}\ .$$

Put

$$L_k(\sigma,\xi) = \prod_{1\leq l\leq r_k} L_{kl}(\sigma,\xi),\quad \underline{\underline{L}}(\sigma,\xi) = \prod_{1\leq k\leq r} L_k(\sigma,\xi)\ .$$

On $L_k(\sigma,\xi)$ we define a norm by $|\varphi_k|^2 = \sum_l |\varphi_{kl}|^2$ if $\varphi_k = (\varphi_{kl})$. Furthermore we put

$${}^{0}\mathcal{D}_k(\sigma,\xi) = \prod_{1\leq l\leq r_k} {}^{0}\mathcal{D}_{kl}(\sigma,\xi),\quad {}^{0}\underline{\underline{\mathcal{D}}}(\sigma,\xi) = \prod_{1\leq k\leq r} {}^{0}\mathcal{D}_k(\sigma,\xi)\ .$$

For $f \in {}^{0}\mathcal{D}_{kl}(\sigma,\xi)$ we define

$$\hat{f}(\Lambda:x) = \int_{A_{kl}} f(xa)\ e^{-(^{y}\Lambda-\varrho)(H(xa))}da \qquad (\Lambda \in \mathfrak{a}^{*}_{k,c})$$

where y,ϱ,H stand respectively for $y_{kl},\varrho_{kl},H_{kl}$ (in Chap.II, 3, we defined the Fourier transform of f as a function on $\mathfrak{a}^{*}_{kl,c}$; here we have transferred it to $\mathfrak{a}^{*}_{k,c}$ by means of y_{kl}). If $f \in {}^{0}\underline{\underline{\mathcal{D}}}(\sigma,\xi)$, $f = (f_k)$, $f_k = (f_{kl})$ with $f_k \in {}^{0}\mathcal{D}_k(\sigma,\xi)$ $f_{kl} \in {}^{0}\mathcal{D}_{kl}(\sigma,\xi)$, then $\hat{f}_k(\Lambda) = (\hat{f}_{kl}(\Lambda))$ is an element of $L_k(\sigma,\xi)$ $(\Lambda \in \mathfrak{a}^{*}_{k,c})$; moreover the functions $E_{f_{kl}}$ are defined and we put

$$E_{f_k} = \sum_{1\leq l\leq r_k} E_{f_{kl}},\quad E_f = \sum_{1\leq k\leq r} E_{f_k}\ .$$

Now we define linear transformations $c_{lk}(s:\Lambda) : L_k(\sigma,\xi) \to L_l(\sigma,\xi)$ for $s \in w(\mathfrak{a}_k,\mathfrak{a}_l)$, $\Lambda \in (\mathfrak{a}^{*}_{k,c})^{-}$. To do this it is sufficient to define $(\psi,c_{lk}(s:\Lambda)\varphi)$ for $\varphi \in L_{kp}(\sigma,\xi)$, $\psi \in L_{lq}(\sigma,\xi)$. Put $(P',A') = (P_{kp},A_{kp})$, $(P'',A'') = (P_{lq},A_{lq})$, $s' = y_{lq}\ s y_{kp}^{-1}$, $\Lambda' = {}^{y_{kp}}\Lambda$. Then $s' \in w(\mathfrak{a}',\mathfrak{a}'')$ and $\Lambda' \in (\mathfrak{a}_c'^{*})^{-}$. Let $c(s':\Lambda')$ be the linear transformation

$${}^{0}L^{2}(M'/\Gamma_{M'},\sigma_{M'},\xi_{P'})\ \cdot\ {}^{0}L^{2}(M''/\Gamma_{M''},\sigma_{M''},\xi_{P''})$$

defined in Theorem 5 (observe that ${}^{s'}\xi_{P'} = \xi_{P''}$). Now by definition

$$(\psi,c_{lk}(s:\Lambda)\varphi) = (\psi,c(s':\Lambda')\varphi)$$

for $\varphi \in L_{kp}(\sigma,\xi)$, $\psi \in L_{lq}(\sigma,\xi)$.

Lemma 40 gives immediately:

Lemma 78. For $f,g \in {}^{o}\underline{\underline{\mathfrak{D}}}(\sigma,\xi)$ we have

$$((E_g,E_f))_{G/\Gamma} = \sum_{1\le k,1\le r} \sum_{s\in w(\mathfrak{a}_k,\mathfrak{a}_1)} \int_{\mathfrak{a}_k^*} (\hat{g}_1(-s\bar{\Lambda}_k),c_{1k}(s:\Lambda_k)\hat{f}_k(\Lambda_k))_1 \, d\Lambda_{k,I}$$

where $\Lambda_{k,R} \in (\mathfrak{a}_k^*)^-$ for $1 \le k \le r$.

And from Lemma 48 one sees at once:

Lemma 79. $c_{1k}(s:\Lambda) = c_{k1}(s^{-1}:-s\bar{\Lambda})^*$

(that is, both sides have the same analytic continuation).

Finally we have:

Lemma 80. $c_{kk}(1:\Lambda) = 1$.

Proof. In the same notation as above we have $(\psi,c_{kk}(1:\Lambda)) = (\psi,c(s':\Lambda')\varphi)$ for $\varphi \in L_{kp}(\sigma,\xi)$, $\psi \in L_{kq}(\sigma,\xi)$, where s' is induced by $y_{kq}\, y_{kp}^{-1}$. If $\Gamma(s') =$ $= \Gamma \cap P''\, y_{kq}\, y_{kp}^{-1} P'$ is not empty, then P' and P'' are conjugate under Γ, so $p = q$. Hence, for $p \neq q$ we have $c(s':\Lambda') = 0$; and for $p = q$ we have $\Gamma(s') = \Gamma \cap P'$ so that $c(s':\Lambda') = 1$.

Lemma 81. If (P,A) and (P',A') are associated cuspidal pairs, then $|\varrho| = |\varrho'|$.

Proof. If (P,A) is mincuspidal, this is clear, because we can choose $y \in G_Q$ such that $(P',A') = {}^{y}(P,A)$. In the general case it is sufficient to prove it for $A = A'$. Then we have $P = MAU$, $P' = MAU'$. Choose a mincuspidal subgroup ${}^*P$ of M and let P_o,P_o' be the corresponding mincuspidal subgroups of G contained respectively in P,P'. Then $P_o = M_oA_oU_o$, $P_o' = M_oA_oU_o'$ with $M_o = {}^*M$, $A_o = {}^*AA$, $U_o = {}^*UU$, $U_o' = {}^*UU'$. Let $2\varrho_o$ be the sum of all positive roots of (P_o,A_o) and $2\varrho_M$ (resp. $2\varrho_P$) the sum of those positive roots which are trivial (resp. are non-trivial) on A. Then $\varrho_o = \varrho_M + \varrho_P$ and $\varrho_M \in \mathfrak{b}^*$, $\varrho_P \in \mathfrak{a}^*$, if $\mathfrak{b} = \mathfrak{a}_o \cap \mathfrak{m}$ (so that $\mathfrak{b}$ is the orthogonal complement of $\mathfrak{a}$ in $\mathfrak{a}_o$). Hence $|\varrho_o|^2 = |\varrho_M|^2 + |\varrho_P|^2$. And for (P_o',A_o) we have $\varrho_o' = \varrho_M + \varrho_{P'}$ (with the same ϱ_M). From $|\varrho_o| = |\varrho_o'|$ it follows now that $|\varrho_p| = |\varrho_{p'}|$, which was to be proved.

Now we apply the results of §3.

By Lemma 81 all numbers $|\varrho_k|$ $(1 \le k \le r)$ are equal; we write simply $|\varrho|$. Fix a real number R with $R > |\varrho|$ and define $\mathfrak{a}_k^*(R)$ and $\mathfrak{a}_{k,C}^*(R)$ as in §3. Let $\mathcal{H}_k(R)$ denote the space of all bounded holomorphic functions h_k on $\mathfrak{a}_{k,C}^*(R)$ with values in $L_k(\sigma,\xi)$ $(\xi \in \mathcal{X}^o(\mathcal{C}))$ such that

$$\int_{\mathfrak{a}_k^*} |h_k(\lambda+i\mu)|^2 d\mu < \infty$$

for any $\lambda \in \mathfrak{a}_k^*(R)$. An element h_k of $\mathcal{H}_k(R)$ has components h_{kp} $(1 \le p \le r_k)$, h_{kp} has its values in $L_{kp}(\sigma,\xi)$. If we put $h'_{kp}({}^{y_{kp}}\Lambda) = h_{kp}(\Lambda)$ for $\Lambda \in \mathfrak{a}_{k,C}^*(R)$, then $\check{E}_{h'_{kp}}$ is defined, call it $\check{E}_{h_k,p}$ and define

$$\check{E}_{h_k} = \sum_{1 \le p \le r_k} \check{E}_{h_k,p} .$$

Furthermore, define

$$\underline{\mathcal{H}}(R) = \prod_{1 \le k \le r} \mathcal{H}_k(R) ,$$

and for $h = (h_k) \in \underline{\mathcal{H}}(R)$ put

$$\check{E}_h = \sum_{1 \le k \le r} \check{E}_{h_k} .$$

Then $\check{E}_h$ is an element of $L^2(G/\Gamma,\sigma)$ which lies in the closure of $E_{\mathcal{C}}(\sigma,\xi)$. For these functions we have the scalar-product formula (cf. Lemma 78):

$$((\check{E}_{h'},\check{E}_h))_{G/\Gamma} = \sum_{1 \le k,1 \le r} \sum_{s \in w(\mathfrak{a}_k,\mathfrak{a}_1)} \int_{\mathfrak{a}_k^*} (h'_1(-s\bar{\Lambda}_k),c_{1k}(s:\Lambda_k)h_k(\Lambda_k))_1 \, d\Lambda_{k,I}$$

where $Re(\Lambda_k) \in (\mathfrak{a}_k^*)^- \cap \mathfrak{a}_k^*(R)$ $\qquad (h,h' \in \underline{\mathcal{H}}(R))$.

The following lemma is analogous to Lemma 72 and is proved in the same way.

Lemma 82. Let $f = (f_k)_{1 \le k \le r}$ be a family of bounded holomorphic functions on $\mathfrak{a}_{k,C}^*(R)$ such that $f_1(s\Lambda) = f_k(\Lambda)$ for $s \in w(\mathfrak{a}_k,\mathfrak{a}_1)$ and $\Lambda \in \mathfrak{a}_{k,C}^*(R)$. There exists a bounded linear transformation T_f on $Cl(E_{\mathcal{C}}(\sigma,\xi))$ such that

$$T_f \check{E}_h = \check{E}_{fh} \qquad \text{for all } h \in \underline{\underline{\mathcal{H}}}(R) .$$

Moreover $\|T_f\| \leq \|f\|_\infty = \max_{1 \leq k \leq r} \|f_k\|_\infty$. Finally $(T_f)^* = T_{f^*}$, where $f^* = (f_k^*)$, $f_k^*(\Lambda) = \overline{f_k(-\bar{\Lambda})}$ $(\Lambda \in \mathfrak{a}^*_{k,C}(\bar{R}))$.

Let ω denote, as before, the Casimir operator of $[\mathfrak{g},\mathfrak{g}]$. If $P \in \mathcal{C}$ and $P = MAU$, then $\mu(\omega) = \omega_{\mathfrak{m}} + \omega_{\mathfrak{a}} - \langle\varrho,\varrho\rangle$, where μ is the canonical homomorphism $\mathfrak{Z} \to \mathfrak{Z}_M \mathfrak{A}$, $\omega_{\mathfrak{m}} \in \mathfrak{Z}_M$, $\omega_{\mathfrak{a}} \in \mathfrak{A}$ such that $\omega_{\mathfrak{a}}(\Lambda) = \langle\Lambda,\Lambda\rangle$. Since ${}^s\mu(z) = \mu(z)$ for any $z \in \mathfrak{Z}$ and $s \in w(\mathfrak{a})$, we have ${}^s\omega_{\mathfrak{m}} = \omega_{\mathfrak{m}}$, so ${}^s\chi(\omega_{\mathfrak{m}}) = \chi(\omega_{\mathfrak{m}})$ for $\chi \in \mathcal{X}_M$. Hence we may define, if ξ is the orbit of χ under $w(\mathfrak{a})$,

$$\xi(\omega:\Lambda) = \chi(\mu(\omega:\Lambda)) = \chi(\omega_{\mathfrak{m}}) + \langle\Lambda,\Lambda\rangle - \langle\varrho,\varrho\rangle .$$

If P' is another element of $\mathcal{C}$ and if A' is a split component of P' and s an element of $w(\mathfrak{a},\mathfrak{a}')$, then $\mu'(\omega) = {}^s\mu(\omega) = {}^s\omega_{\mathfrak{m}} + {}^s\omega_{\mathfrak{a}} - \langle\varrho,\varrho\rangle$ and $\xi'(\omega:{}^s\Lambda) = {}^s\chi(\mu'(\omega:{}^s\Lambda)) = \chi(\omega_{\mathfrak{m}}) + \langle\Lambda,\Lambda\rangle - \langle\varrho,\varrho\rangle$. Hence, returning to our previous notations and applying the above to the groups P_k, we may define for $1 \leq k \leq r$, $\xi \in \mathcal{X}^o(\mathcal{C})$,

$$\xi_k(\omega:\Lambda) = \chi(\mu_k(\omega:\Lambda)) = \langle\Lambda,\Lambda\rangle + c(\xi) \qquad (\Lambda \in \mathfrak{a}^*_{k,C}) ,$$

where χ is any element in the component of ξ in $\mathcal{X}^o(P_k)$; $c(\xi)$ is a number which depends only on ξ and which is real if $\underline{\underline{L}}(\sigma,\xi) \neq 0$.

Let $\underline{\underline{\mathcal{H}}}^o(R)$ denote the subspace of $\underline{\underline{\mathcal{H}}}(R)$ consisting of the elements $h = (h_k) \in \underline{\underline{\mathcal{H}}}(R)$ such that for every k and every polynomial function p on $\mathfrak{a}^*_{k,C}$, the function ph_k is bounded on $\mathfrak{a}^*_{k,C}(R)$. For $h \in \underline{\underline{\mathcal{H}}}^o(R)$ and $\xi \in \mathcal{X}^o(\mathcal{C})$ we define an element $\xi(\omega)h$ in $\underline{\underline{\mathcal{H}}}(R)$ by

$$(\xi(\omega)h)_k(\Lambda) = \xi_k(\omega:\Lambda)h_k(\Lambda) \qquad (\Lambda \in \mathfrak{a}^*_{k,C}(R)).$$

By Lemma 77 we have now

$$\omega \check{E}_h = \check{E}_{\xi(\omega)h} \qquad \text{for } h \in \underline{\underline{\mathcal{H}}}^o(R).$$

Put $\omega_o = \omega - c(\xi)$. Then $\omega_o\check{E}_h = \check{E}_{h'}$, where $h_k'(\Lambda) = \langle\Lambda,\Lambda\rangle h_k(\Lambda)$.

Fix a complex number ζ with $\mathrm{Re}(\zeta) > R^2$. Put $f_\zeta = (f_k)_{1\le k\le r}$, $f_k(\Lambda) = (\zeta - \langle\Lambda,\Lambda\rangle)^{-1}$ for $\Lambda \in \mathfrak{a}^*_{k,c}(R)$. The functions f_k are bounded, so we can apply Lemma 82, which gives us a bounded operator $T_\zeta = T_{f_\zeta}$ such that $T_\zeta \check{E}_h = \check{E}_{f_\zeta h}$ for all $h \in \underline{\underline{\mathcal{H}}}(R)$. From the scalar-product formula one sees that $((\check{E}_{h'}, T_\zeta \check{E}_h))_{G/\Gamma}$ is a holomorphic function of ζ for $\mathrm{Re}(\zeta) > R^2$ (h and h' arbitrary in $\underline{\underline{\mathcal{H}}}(R)$). Let R_ζ denote the resolvent of ω_o. Then, for $h \in \underline{\underline{\mathcal{H}}}^o(R)$ we have

$$R_\zeta \check{E}_h = T_\zeta \check{E}_h \qquad (\mathrm{Re}(\zeta) > R^2),$$

since $(\zeta - \omega_o)T_\zeta \check{E}_h = (\zeta - \omega_o)\check{E}_{f_\zeta h} = \check{E}_h$. Now $((\check{E}_{h'}, R_\zeta \check{E}_h))_{G/\Gamma}$ is holomorphic for $\zeta \in \mathbb{C}$, $\zeta \notin \mathbb{R}$, so we get finally:

<u>Lemma 83</u>. Let $h, h' \in \underline{\underline{\mathcal{H}}}^o(R)$. Then, if T_ζ is defined as above, the scalar product $((\check{E}_{h'}, T_\zeta \check{E}_h))_{G/\Gamma}$ is a holomorphic function of ζ for all ζ except possibly when ζ is real and $\le R^2$.

§5. Analytic continuation of $\underline{c}(z)$ and $E(z:\varphi)$: first step.

We use the notations of §4 and suppose now that rank $\mathcal{C} = 1$, i.e. the groups in the class $\mathcal{C}$ are maximal cuspidal subgroups of G. If $P \in \mathcal{C}$ and $P = MAU$, then any group in $\mathcal{C}$ is conjugate to P or to the opposite group $P^- = MAU^-$, and P and P^- are conjugate if and only if $-1 \in w(\mathfrak{a})$. So the number r of §4 is 1 or 2; if $r = 1$, then $w(\mathfrak{a}_1) = \{\pm 1\}$ and if $r = 2$, then $w(\mathfrak{a}_k) = \{1\}$ for $k = 1,2$. In either case, put $\lambda_k = \alpha_k/|\alpha_k|$, where α_k is the simple root of (P_k, A_k) $(1 \le k \le r)$. Then $\varrho_k = |\varrho|\lambda_k$.

If $r = 1$, define $\underline{c}(z) = c_{11}(-1:z\lambda_1)$ for $\mathrm{Re}(z) < -|\varrho|$. This is a linear transformation $\underline{\underline{L}} \to \underline{\underline{L}}$, where $\underline{\underline{L}} = \underline{\underline{L}}(\sigma,\xi) = L_1(\sigma,\xi)$.

If $r = 2$, we have $\underline{\underline{L}} = \underline{\underline{L}}(\sigma,\xi) = L_1(\sigma,\xi) \times L_2(\sigma,\xi)$. Define in this case $\underline{c}(z)$ as the linear transformation $\underline{\underline{L}} \to \underline{\underline{L}}$ given by the matrix

$$\begin{pmatrix} 0 & c_{12}(s^{-1}:z\lambda_2) \\ c_{21}(s:z\lambda_1) & 0 \end{pmatrix}$$

where s is the unique element of $w(\mathfrak{a}_1, \mathfrak{a}_2)$ and $\operatorname{Re}(z) < -|\varrho|$.

In both cases we have $\underline{\underline{c}}(z)^* = \underline{\underline{c}}(\bar{z})$ (Lemma 79) and in both cases we have the scalar-product formula

$$((\check{E}_{h'}, \check{E}_h))_{G/\Gamma} = \frac{1}{2\pi i} \int_{c-i\infty}^{c+i\infty} \{(h'(-\bar{z}), h(z))_{\underline{\underline{L}}} + (h'(\bar{z}), \underline{\underline{c}}(z)h(z))_{\underline{\underline{L}}}\} dz$$

for $-R < c < -|\varrho|$, h and $h' \in \underline{\underline{\mathcal{H}}}(R)$; $h(z)$ stands for $h(z\lambda_1)$ in the case $r = 1$ and for $\begin{pmatrix} h_1(z\lambda_1) \\ h_2(z\lambda_2) \end{pmatrix}$ in the case $r = 2$.

Put $\lambda_{kl} = \alpha_{kl}/|\alpha_{kl}|$, where α_{kl} is the simple root of (P_{kl}, A_{kl}), and $t_{kl}(x) = \lambda_{kl}(H_{P_{kl}|A_{kl}}(x))$.

Let D denote the complement of the real interval $[-|\varrho|, 0[$ in the half-plane $\operatorname{Re}(z) < 0$.

Lemma 84. For any $\varphi, \psi \in \underline{\underline{L}}$, $(\psi, \underline{\underline{c}}(z)\varphi)_{\underline{\underline{L}}}$ is a holomorphic function of z for $z \in D$.

Proof. Consider the functions h and h' defined by $h(z) = e^{z^2}\varphi$, $h'(z) = e^{z^2}\psi$. They belong to $\underline{\underline{\mathcal{H}}}^{\circ}(R)$ for any R. Let T_ζ be the operator on $\mathrm{Cl}(E_{\mathcal{C}}(\sigma,\xi))$ defined at the end of §4. Then $T_{z^2}\check{E}_h = \check{E}_{h''}$ where $h''(\zeta) = (z^2 - \zeta^2)^{-1}h(\zeta)$ and the scalar-product formula gives

$$((\check{E}_{h'}, T_{z^2}\check{E}_h))_{G/\Gamma} = \frac{1}{2\pi i} \int_{c-i\infty}^{c+i\infty} \frac{e^{2\zeta^2}}{z^2-\zeta^2} \{(\psi,\varphi)_{\underline{\underline{L}}} + (\psi, \underline{\underline{c}}(\zeta)\varphi)_{\underline{\underline{L}}}\} d\zeta$$

for $\operatorname{Re}(z) < 0$, $\operatorname{Re}(z^2) > c^2$, $c < -|\varrho|$. By shifting the integration to a line $\operatorname{Re}(\zeta) = c_1$ with $c_1 < c$ we get

$$((\check{E}_{h'}, T_{z^2}\check{E}_h))_{G/\Gamma} = \frac{1}{2\pi i} \int_{c_1-i\infty}^{c_1+i\infty} \cdots - \frac{e^{2z^2}}{2z} \{(\psi,\varphi)_{\underline{\underline{L}}} + (\psi, \underline{\underline{c}}(z)\varphi)_{\underline{\underline{L}}}\}$$

if $c_1 < \operatorname{Re}(z) < 0$, $\operatorname{Re}(z^2) > c^2$. By Lemma 83 the left hand side is holomorphic for $z \in D$; clearly the integral over $\operatorname{Re}(\zeta) = c_1$ is holomorphic for $c_1 < \operatorname{Re}(z) < 0$. Hence $(\psi, \underline{\underline{c}}(z)\varphi)_{\underline{\underline{L}}}$ is holomorphic for $z \in D$.

We are going to apply the Maass-Selberg relation.

Let $\mathcal{P}_1$ denote the set of all cuspidal subgroups of G of rank 1 ; so $\mathcal{P}_1 \supset \mathcal{C}$. Fix a set of representatives $\mathcal{P}_1^o$ for $\mathcal{P}_1/\Gamma$ which contains the set $\mathcal{C}^o = \{P_{kl}\}$ of representatives for $\mathcal{C}/\Gamma$. We change a little the notations of §1: the partition of G/Γ introduced there is now

$$G/\Gamma = {}^c\tilde{S}(t) \cup \bigcup_{P \in \mathcal{P}_1^o} \tilde{S}_P(t) \ ;$$

π_P (resp. π_P') denotes the canonical map $G/\Gamma \cap P \to G/\Gamma$ (resp. $G \to G/\Gamma \cap P$), etc. We write $\tilde{S}_{kl}'(t)$, π_{kl}' , etc., instead of $\tilde{S}_{P_{kl}}'(t)$, $\pi_{P_{kl}}'$, etc. If f is a function belonging to $\mathcal{A}_1(G/\Gamma,\sigma)$, the <u>modified function</u> $\tilde{f}$ is defined by

$$\tilde{f} = f \quad \text{on} \quad {}^c\tilde{S}(t), \quad \tilde{f} = f - f_P \quad \text{on} \quad \tilde{S}_P'(t) \qquad (P \in \mathcal{P}_1^o)$$

(recall that π_P defines a homeomorphism of $\tilde{S}_P'(t)$ on $\tilde{S}_P(t)$).

Put $\underline{\underline{\Phi}} = \underline{\underline{L}} \times \underline{\underline{L}}$, and for any complex number $z \neq 0$ let $L(z)$ be the space of all functions f in $\mathcal{A}_1(G/\Gamma,\sigma)$ for which there exists an element $h = (\varphi,\psi) \in \underline{\underline{\Phi}}$ such that

1) $f_P = o$ if $P \in \mathcal{P}_1^o$, $P \notin \mathcal{C}^o$,

2) $f_{P_{kl}}(x) = \varphi_{kl}(x)\, e^{(z-|\varrho|)t_{kl}(x)} + \psi_{kl}(x)\, e^{-(z+|\varrho|)t_{kl}(x)}$ (all k,l).

This is a special case of the first definition of Chap. III,§1 (take as a base for B_{kl} the elements $\frac{1}{2}(1 \pm z^{-1}H_{kl})$ where $H_{kl} \in \mathfrak{a}_{kl}$, $\alpha_{kl}(H_{kl}) > o$, $|H_{kl}| = 1$; $\Phi_P = o$ if $P \in \mathcal{P}_1^o$, $P \notin \mathcal{C}^o$). A more complete notation for $L(z)$ is $\mathcal{A}_1(G/\Gamma,\sigma,\underline{\underline{\Phi}},z)$. Clearly $L(z) = L(-z)$. Using the notations of Chap.III,§1 we have now a bijection $h \longmapsto f(h:z)$ of a subspace $\underline{\underline{\Phi}}(z)$ of $\underline{\underline{\Phi}}$ on $L(z)$.

Consider the Eisenstein series $E(z:\varphi) = \sum_{k,l} E(z:\varphi_{kl})$ where $\varphi = (\varphi_{kl}) \in \underline{\underline{L}}$ and $E(z:\varphi_{kl})$ is defined, in accordance with Chap.II,§2, by

$$E(z:\varphi_{kl}:x) = \sum_{\Gamma/\Gamma \cap P_{kl}} \varphi_{kl}(x\gamma)\, e^{(z-|\varrho|)t_{kl}(x\gamma)} ,$$

the series being absolutely convergent for $Re(z) < -|\varrho|$. If $f = E(z:\varphi)$, then

$f \in \mathcal{A}_1(G/\Gamma,\sigma)$ and $f_P = 0$ if $P \in \mathcal{P}_1$, $P \notin \mathcal{C}$ (Chap.II,§4). Moreover

$$f_{P_{kl}}(x) = \varphi_{kl}(x)e^{(z-|\varrho|)t_{kl}(x)} + (\underline{\underline{c}}(z)\varphi)_{kl}(x)e^{-(z+|\varrho|)t_{kl}(x)}$$

(Theorem 5). Hence $E(z:\varphi) \in L(z)$ and $E(z:\varphi) = f(h:z)$ with $h = (\varphi,\underline{\underline{c}}(z)\varphi) \in \underline{\underline{L}} \times \underline{\underline{L}}$.

If ω is again the Casimir operator of $[\mathfrak{g},\mathfrak{g}]$ and if $\omega_o = \omega - c(\xi)$ as at the end of §4, then for $f \in L(z)$ we have $\omega_o f = z^2 f$ (because $\omega_o f \in \mathcal{A}_1(G/\Gamma,\sigma)$ and $(\omega_o f)_P = \omega_o f_P = z^2 f_P$ for $P \in \mathcal{P}_1^o$). Hence, if $f \in L(z)$ and $g \in L(\zeta)$, then $[f,g] = (\omega f,g) - (f,\omega g) = (\bar{z}^2 - \zeta^2)(f,g)$ and the Maass-Selberg relation gives

$$(\bar{z}^2 - \zeta^2)((\tilde{f},\tilde{g}))_{G/\Gamma} = \sum_{k,l}(\bar{z}^2 - \zeta^2)\int_{G'_{kl}(t)\cap{}^c\bar{S}'_{kl}(t)}(f_{P_{kl}},g_{P_{kl}})dx - \sum_{k,l} J_{P_{kl}}(f,g:t')$$

where $\tilde{f}$ and $\tilde{g}$ are the modified functions and where t' stands for $|\alpha_{kl}|^{-1}t$. One computes trivially from the definition of $J_P(f,g:t)$ that

$$\sum_{k,l} J_{P_{kl}}(f,g:t') = (\bar{z} - \zeta)\{(\varphi,\varphi')_{\underline{\underline{L}}}\, e^{t'(\bar{z}+\zeta)} - (\psi,\psi')_{\underline{\underline{L}}}\, e^{-t'(\bar{z}+\zeta)}\} +$$

$$+ (\bar{z} + \zeta)\{(\varphi,\psi')_{\underline{\underline{L}}}\, e^{t'(\bar{z}-\zeta)} - (\psi,\varphi')_{\underline{\underline{L}}}\, e^{-t'(\bar{z}-\zeta)}\}$$

if $f = f(h:z)$, $h = (\varphi,\psi)$, $g = f(h':\zeta)$, $h' = (\varphi',\psi')$.

<u>Lemma 85</u>. a) For $z \neq 0$ one has the inequalities

$$\dim \underline{\underline{\Phi}}(z) + \dim \underline{\underline{\Phi}}(\bar{z}) \leq \dim \underline{\underline{\Phi}} = 2 \dim \underline{\underline{L}},$$

$$\dim \underline{\underline{\Phi}}(z) + \dim \underline{\underline{\Phi}}(-\bar{z}) \leq \dim \underline{\underline{\Phi}}.$$

b) If $h = (\varphi,\psi) \in \underline{\underline{\Phi}}(z) \cap \underline{\underline{\Phi}}(-\bar{z})$, then $|\varphi|_{\underline{\underline{L}}} = |\psi|_{\underline{\underline{L}}}$.

<u>Proof</u>. It follows from the Maass-Selberg relation above that $(\varphi,\psi')_{\underline{\underline{L}}} = (\psi,\varphi')_{\underline{\underline{L}}}$ if $\zeta = \bar{z}$ and that $(\varphi,\varphi')_{\underline{\underline{L}}} = (\psi,\psi')_{\underline{\underline{L}}}$ if $\zeta = -\bar{z}$. The latter gives already the assertion b). Considering the linear transformations

$$j : (\varphi,\psi) \longmapsto (\varphi,-\psi) \quad \text{and} \quad j' : (\varphi,\psi) \longmapsto (\psi,-\varphi)$$

of $\underline{\underline{\Phi}}$ one sees that $\underline{\underline{\Phi}}(z)$ is orthogonal to $j(\underline{\underline{\Phi}}(-\bar{z}))$ and to $j'(\underline{\underline{\Phi}}(\bar{z}))$, hence a).

Take now in the Maass-Selberg relation $f = E(z:\varphi)$, $g = E(\zeta:\psi)$ with $\varphi,\psi \in \underline{\underline{L}}$, $\mathrm{Re}(z) < -|\varrho|$, $\mathrm{Re}(\zeta) < -|\varrho|$. This gives

$$((\tilde{E}(z:\varphi),\tilde{E}(\zeta:\psi)))_{G/\Gamma} = \Omega_{\varphi,\psi}(\bar{z}:\zeta)$$

where

$$\Omega_{\varphi,\psi}(\bar{z}:\zeta) = \sum_{k,l} I_{kl}(\bar{z}:\zeta) + \Omega_o(\bar{z}:\zeta) ,$$

$$I_{kl}(\bar{z}:\zeta) = \int_{G'_{kl}(t)\cap{}^c\tilde{S}'_{kl}(t)} (f_{P_{kl}}, g_{P_{kl}})\, dx ,$$

$$\Omega_o(\bar{z}:\zeta) = -\frac{1}{\bar{z}+\zeta}\{(\varphi,\psi)_{\underline{\underline{L}}}\, e^{t'(\bar{z}+\zeta)} - (\underline{\underline{c}}(z)\varphi,\underline{\underline{c}}(\zeta)\psi)_{\underline{\underline{L}}}\, e^{-t'(\bar{z}+\zeta)}\}$$

$$-\frac{1}{\bar{z}-\zeta}\{(\varphi,\underline{\underline{c}}(\zeta)\psi)_{\underline{\underline{L}}}\, e^{t'(\bar{z}-\zeta)} - (\underline{\underline{c}}(z)\varphi,\psi)_{\underline{\underline{L}}}\, e^{-t'(\bar{z}-\zeta)}\} .$$

From Lemma 84 and the fact that $\underline{\underline{c}}(z)^* = \underline{\underline{c}}(\bar{z})$ we see that $\Omega_o(z:\zeta)$ is holomorphic on $D \times D$. From the proof of the Maass-Selberg relation (which we did not give) one can see that the integral

$$\int_{G'_{kl}(t)\cap{}^c S'_{kl}(t)} |f_{P_{kl}}|\,|g_{P_{kl}}|\, dx$$

converges uniformly on every compact subset of $D \times D$, so $I_{kl}(z:\zeta)$ is holomorphic on $D \times D$. Hence $\Omega_{\varphi,\psi}(z:\zeta)$ is holomorphic on $D \times D$.

Moreover, it can be proved that $\Omega_{\varphi,\varphi}(\bar{z}:z) \leq c\,\|\varphi\|^2_{\underline{\underline{L}}}$ for all $\varphi \in \underline{\underline{L}}$ if z stays in a compact subset of D.

Lemma 86. Fix $\varphi \in \underline{\underline{L}}$. The mapping $z \longmapsto \tilde{E}(z:\varphi)$ of the set $D_1 = \{z : \mathrm{Re}(z) < -|\varrho|\}$ into $L^2(G/\Gamma,\sigma)$ is holomorphic.

Proof. It is clear that the mapping is continuous. Fix $z_o \in D_1$ and define

$$F(z) = \frac{1}{2\pi i}\int_C \frac{\tilde{E}(\zeta:\varphi)}{\zeta - z}\, d\zeta$$

where C is a small circle around z_o. Then $F(z) \in L^2(G/\Gamma,\sigma)$ for z in a neighbourhood of z_o and

$$((\alpha,F(z)))_{G/\Gamma} = \frac{1}{2\pi i}\int_C \frac{((\alpha,\tilde{E}(\zeta:\varphi)))_{G/\Gamma}}{\zeta-z}\,d\zeta = ((\alpha,\tilde{E}(z:\varphi)))_{G/\Gamma}$$

for any $\alpha \in C_c(G/\Gamma,\sigma)$. So $F(z) = \tilde{E}(z:\varphi)$.

Lemma 87. The mapping $z \longmapsto \tilde{E}(z:\varphi)$ extends to a holomorphic mapping $D \to L^2(G/\Gamma,\sigma)$.

Proof. We shall prove the following. Let $z_o \in D$ and $a > 0$ such that $|z-z_o| < a \to z \in D$. Assume that there exists a neighbourhood U of z_o such that $\tilde{E}(z:\varphi)$ extends holomorphically to U for every $\varphi \in \underline{\underline{L}}$ and that $((\tilde{E}(z:\varphi), \tilde{E}(\zeta:\psi)))_{G/\Gamma} = \Omega_{\varphi,\psi}(\bar{z}:\zeta)$ for $\varphi,\psi \in \underline{\underline{L}}$, $z, \zeta \in U$. Then $\tilde{E}(z:\varphi)$ extends holomorphically to the set $\{z:|z-z_o| < a\}$.

Put $E_n(\varphi) = \left(\frac{d^n}{dz^n}\tilde{E}(z:\varphi)\right)_{z=z_o}$. We have to prove that $\sum_{n\geq o} \|E_n(\varphi)\|\frac{b^n}{n!}$ converges for $o < b < a$. By assumption

$$\|E_n(\varphi)\|^2 = ((E_n(\varphi),E_n(\varphi))) = \left(\frac{\partial^{2n}}{\partial\zeta^n\partial z^n}\,\Omega_{\varphi,\varphi}(\zeta:z)\right)_{\zeta=\bar{z}_o,\ z=z_o}.$$

Put $c_{mn} = \left(\frac{\partial^{m+n}}{\partial\zeta^m\partial z^n}\,\Omega_{\varphi,\varphi}(\zeta:z)\right)_{\zeta=\bar{z}_o,\ z=z_o}$. Then $\Sigma|c_{mm}|(\frac{b^m}{m!})^2 < \infty$ for $b < a$ since $\Omega_{\varphi,\varphi}$ is holomorphic on $D\times D$. So $\|E_m(\varphi)\|\frac{b^m}{m!} = |c_{mm}|^{1/2}\frac{b^m}{m!}$ tends to zero for $m \to \infty$ $(b < a)$, which implies the convergence of $\Sigma\|E_m(\varphi)\|\frac{b^m}{m!}$ for $b < a$.

Let β_{kl} be the characteristic function of $(\pi'_{kl})^{-1}(\tilde{S}'_{kl}(t))$. Define $f'_{kl}(z:\varphi:x) = \beta_{kl}(x)\varphi_{kl}(x)\,e^{(z-|\varrho|)t_{kl}(x)}$ $(\varphi \in \underline{\underline{L}},\ z \in C,\ x \in G)$, so that $f'_{kl}(z:\varphi)$ is right-invariant under $\Gamma \cap P_{kl}$; and define the function $f_{kl}(z:\varphi)$ by

$$\begin{cases} f_{kl}(z:\varphi:x) = 0 \quad \text{if} \quad \pi(x) \notin \tilde{S}_{kl}(t), \\ f_{kl}(z:\varphi:x) = f'_{kl}(z:\varphi:x) \quad \text{if} \quad \pi'_{kl}(x) \in \tilde{S}'_{kl}(t), \\ f_{kl}(z:\varphi) \quad \text{is right-invariant under } \Gamma . \end{cases}$$

Then we have

$$E(z:\varphi:x) = \tilde{E}(z:\varphi:x) + \sum_{k,l}(f_{kl}(z:\varphi:x) + f_{kl}(-z:\underline{\underline{c}}(z)\varphi:x))$$

for $Re(z) < -|\varrho|$. The terms in the sum over k,l are holomorphic functions of z for

$z \in D$ (Lemma 84) and by Lemma 87 $\tilde{E}(z:\varphi)$ is defined for $z \in D$ as an element of $L^2(G/\Gamma,\sigma)$, so we can define the distribution $E(z:\varphi)$ for $z \in D$:

$$E(z:\varphi:\beta) = \int_G \tilde{E}(z:\varphi:x)\beta(x)dx + \sum_{k,l}\{\int_G f_{kl}(z:\varphi:x)\beta(x)dx + \int_G f_{kl}(-z:\underline{\underline{c}}(z)\varphi:x)\beta(x)dx\}$$

for $\beta \in C_c^\infty(G)$. Now we need a few lemmas.

Lemma 88. Let $\mathfrak{S}$ and $\mathfrak{S}'$ be Siegel domains in G with respect to mincuspidal pairs (P,A) and (P',A') respectively. Let $y \in G_{\mathbb{Q}}$ and let Γ_o be the set of all $\gamma \in \Gamma$ such that $\mathfrak{S} \cap \mathfrak{S}'y\gamma$ is not empty. Then $\Gamma_o(\Gamma \cap U)/\Gamma \cap U$ is finite.

Proof. Choose $\xi \in G_{\mathbb{Q}}$ such that $(P,A) = {}^\xi(P',A')$. If $\mathfrak{S} = KA_t\omega$, put $\mathfrak{S}'_o = K^{\xi^{-1}}(A_t\omega)$. There is a Siegel domain $\mathfrak{S}'_1$ with respect to (P',A') such that $K\xi\mathfrak{S}'_o \subset \mathfrak{S}'_1 U' = \mathfrak{S}'_1(\xi^{-1}\Gamma \cap U')$. Now $\mathfrak{S} \subset K\xi\mathfrak{S}'_o\xi^{-1} \subset \mathfrak{S}'_1\xi^{-1}(\Gamma \cap U)$. If $\gamma \in \Gamma_o$, we can choose $\gamma_1 \in \gamma(\Gamma \cap U)$ such that $\mathfrak{S}'_1\xi^{-1} \cap \mathfrak{S}'y\gamma_1$ is not empty; then $\mathfrak{S}'_1\xi^{-1}y^{-1} \cap \mathfrak{S}'{}^y\gamma_1$ is not empty. The set of all $\beta \in {}^y\Gamma$ such that $\mathfrak{S}'_1\xi^{-1}y^{-1} \cap \mathfrak{S}'\beta$ is not empty is finite, so γ_1 belongs to a finite set and the lemma is proved.

In the following lemma we use the notation of §1.

Lemma 89. Let $\mathfrak{S}$ be a Siegel domain with respect to a mincuspidal pair (P_o,A_o) and Ξ a finite subset of $G_{\mathbb{Q}}$. If Γ_i is the set of all $\gamma \in \Gamma$ such that $\mathfrak{S}\Xi\gamma \cap \bigcup_j \mathfrak{S}_{ij}B^o_{ij}$ is not empty, then $\Gamma_i(\Gamma \cap P_i)/\Gamma \cap P_i$ is finite.

Proof. Let $\Gamma_{ij}(\xi)$ denote the set of all $\gamma \in \Gamma$ such that $\mathfrak{S}\xi\gamma \cap \mathfrak{S}_{ij}B^o_{ij}$ is not empty ($\xi \in \Xi$). Since $\mathfrak{S}_{ij}B^o_{ij}$ is contained in a Siegel domain with respect to (P_{ij},A_{ij}), Lemma 88 implies that $\Gamma_{ij}(\xi)(\Gamma \cap U_{ij})/\Gamma \cap U_{ij}$ is finite. Now $\Gamma \cap U_{ij} \subset \Gamma \cap P_i$, and $\Gamma_i(\Gamma \cap P_i)/\Gamma \cap P_i = \bigcup_{j,\xi} \Gamma_{ij}(\xi)(\Gamma \cap P_i)/\Gamma \cap P_i$ is finite.

Lemma 90. Let (P,A) be a cuspidal pair of G, let $y \in G_{\mathbb{Q}}$ and $(P',A') = {}^y(P,A)$. Let ω^* be a compact subset of $\mathfrak{a}^*$. Then

$$\sup_{x\in G,\lambda\in\omega^*} |{}^y\lambda(H'(x)) - \lambda(H(xy))| < \infty ;$$

here $H = H_{P|A}$, $H' = H_{P'|A'}$.

Proof. If $x = kp$, $k \in K$, $p \in P$, then $^yx = {}^yk^yp = {}^ykp'$, $p' \in P'$ and $H'(^yx) = H'(^yk) + H'(p') = H'(^yk) + {}^yH(x)$; $H(yx) = H(yk) + H(x)$. So $H'(^yx) - {}^yH(yx)$ is bounded for $x \in G$, $H'(x) - {}^yH(xy)$ is bounded, and ${}^y\lambda(H'(x)) - \lambda(H(xy)) = {}^y\lambda(H'(x) - {}^yH(xy))$.

After these preparations we can estimate the functions f_{kl} introduced above.

Lemma 91. Suppose z varies in such a way that $|\mathrm{Re}(z)|$ remains bounded. Then we can choose a constant c such that

$$|f_{kl}(z:\varphi:x)| \le c\, \|\varphi\|_{\underline{\underline{L}}}\, \|x\|^N$$

for some N, all $\varphi \in \underline{\underline{L}}$, all $x \in G$.

Proof. Choose a Siegel domain $\mathfrak{S}$ with respect to a mincuspidal pair (P_o, A_o) and a finite subset Ξ of $G_{\mathbb{Q}}$ such that $G = \mathfrak{S}\,\Xi\,\Gamma$. Denote by Γ_{kl} the set of all $\gamma \in \Gamma$ such that $\pi'_{kl}(\mathfrak{S}\,\Xi\,\gamma) \cap \tilde{S}'_{kl}(t)$ is not empty. By Lemma 89, $\Gamma_{kl}/\Gamma \cap P_{kl}$ is finite; choose a set Γ^o_{kl} of representatives. For $x \in \mathfrak{S}\Xi$ we have

$$|f_{kl}(z:\varphi:x)| \le \sum_{\Gamma/\Gamma\cap P_{kl}} |f'_{kl}(z:\varphi:x\gamma)| = \sum_{\Gamma_{kl}/\Gamma\cap P_{kl}} |f'_{kl}(z:\varphi:x\gamma)| \le$$

$$\le \sum_{\Gamma^o_{kl}} |\varphi_{kl}(x\gamma)|\, e^{(u-|\varrho|)t_{kl}(x\gamma)} \le c_o\, \|\varphi\|_{\underline{\underline{L}}} \sum_{\Gamma^o_{kl}} e^{(u-|\varrho|)t_{kl}(x\gamma)}$$

where $u = \mathrm{Re}(z)$. Fix $y_{kl} \in G_{\mathbb{Q}}$ such that $(P'_{kl}, A'_{kl}) = {}^{y_{kl}}(P_{kl}, A_{kl}) \succ (P_o, A_o)$. By Lemma 90, $|(u-|\varrho|)t_{kl}(x) - (u-|\varrho|)t'_{kl}(xy^{-1}_{kl})|$ is bounded for $x \in G$, u bounded; here t'_{kl} is defined by $t'_{kl}(x) = {}^{y_{kl}}\lambda_{kl}(H'_{kl}(x))$. So

$$|f_{kl}(z:\varphi:x\xi)| \le c_1\, \|\varphi\|_{\underline{\underline{L}}} \sum_{\Gamma^o_{kl}} e^{(u-|\varrho|)t'_{kl}(x\xi\gamma y^{-1}_{kl})} \qquad (x \in \mathfrak{S},\ \xi \in \Xi,\ u \text{ bounded}).$$

Now $e^{(u-|\varrho|)t'_{kl}(x\xi\gamma y^{-1}_{kl})} \le c_2\, \|x\|^N$ and $\|x\| \le c_3\, \|x\xi\|$, so

$$|f_{kl}(z:\varphi:x)| \le c_4\, \|\varphi\|_{\underline{\underline{L}}}\, \|x\|^N \text{ for } x \in \mathfrak{S}\Xi, \text{ hence } |f_{kl}(z:\varphi:x)| \le c\, \|\varphi\|_{\underline{\underline{L}}}\, \|x\|^N .$$

for $x \in G$.

Lemma 92. Given compact subsets D_o and Ω of D and G respectively, there is a constant c such that

$$|E(z:\varphi:\beta)| \leq c \, \|\beta\|_\infty \, \|\varphi\|_{\underline{\underline{L}}}$$

for $z \in D_o$, $\varphi \in \underline{\underline{L}}$, $\beta \in C_c^\infty(G)$ with $\mathrm{Supp}(\beta) \subset \Omega$. $E(z:\varphi:\beta)$ is holomorphic in z for $z \in D$; it is continuous in (z,β) for $z \in D$, $\beta \in C_c^\infty(G)$.

<u>Proof</u>. $|\int_G \tilde{E}(z:\varphi:x)\beta(x)dx| \leq \|\beta\|_\infty \int_\Omega |\tilde{E}(z:\varphi:x)|dx$ and, if δ is the characteristic function of Ω, we have

$$\int_\Omega |E(z:\varphi:x)|dx = \int_{G/\Gamma} \sum_\Gamma \delta(x\gamma)|\tilde{E}(z:\varphi:x)|dx \leq c_1 \int_{G/\Gamma} |\tilde{E}(z:\varphi:x)|dx \leq$$

$\leq c_2 \|\tilde{E}(z:\varphi)\|_2 \leq c_3 \|\varphi\|_{\underline{\underline{L}}}$ for $z \in D_o$. And $|\int_G f_{kl}(z:\varphi:x)\beta(x)dx| \leq$

$\leq \|\beta\|_\infty \int_\Omega c \, \|\varphi\|_{\underline{\underline{L}}} \, \|x\|^N dx$ (Lemma 91); and also by Lemma 91,

$|f_{kl}(-z:\underline{\underline{c}}(z)\varphi:x)| \leq c \, \|\underline{\underline{c}}(z)\varphi\|_{\underline{\underline{L}}} \, \|x\|^N \leq c' \, \|\varphi\|_{\underline{\underline{L}}} \, \|x\|^N$ for $z \in D_o$. This gives the estimation for $E(z:\varphi:\beta)$; the rest follows easily.

We know that

$$\alpha * E(z:\varphi) = E(z:\pi(z:\alpha)\varphi) \qquad (\alpha \in I_c^\infty(G),\ \mathrm{Re}(z) < -|\rho|),$$

where $\pi(z)$ is a representation of $I_c^\infty(G)$ on $\underline{\underline{L}}$ (cf. Chap.II,§2: $\beta * \varphi_\Lambda = (\pi_\Lambda(\beta)\varphi)_\Lambda$). By Lemma 92 this equality holds as an equality of distributions also for $z \in D$. Fix $z_o \in D$ and choose $\alpha \in I_c^\infty(G)$ such that $\pi(z_o:\alpha) = 1$. Choose a compact neighbourhood D_o of z_o in D such that $\det \pi(z:\alpha) \neq 0$ for $z \in D_o$. Then $E(z:\varphi) = \alpha * E(z:\pi(z:\alpha)^{-1}\varphi)$ for $z \in D_o$. The right hand side in this formula is a C^∞ function on G, so the distribution $E(z:\varphi)$ is in fact a C^∞ function. If $g \in \mathfrak{g}$, then $E(z:\varphi:g_\iota x) = \int \alpha(g_\iota xy^{-1})E(z:\pi(z:\alpha)^{-1}\varphi:y)dy = E(z:\pi(z:\alpha)^{-1}\varphi:\beta_x)$, where $\beta_x(y) = \alpha(g_\iota xy^{-1})$. Since $x \longmapsto \beta_x$ is a continuous mapping $G \to C_c^\infty(G)$, $E(z:\varphi:g_\iota x)$ is a continuous function of (z,x) on $D \times G$. Finally we estimate $E(z:\varphi:x)$, as follows.

$$\int\alpha(xy^{-1})E(z:\varphi:y)dy = \int\alpha(xy^{-1})\tilde{E}(z:\varphi:y)dy + \sum_{k,l}\ldots,$$

$$|\int\alpha(xy^{-1})\tilde{E}(z:\varphi:y)dy| \leq \int_{G/\Gamma}\sum_{\Gamma}|\alpha(x\gamma y^{-1})|\,|\tilde{E}(z:\varphi:y)|dy \leq$$

$$\leq c_o\|x\|^N\,\|\tilde{E}(z:\varphi)\|_2 \leq c_1\|x\|^N\,\|\varphi\|_{\underline{\underline{L}}} \quad \text{for } x \in G,\ z \in D_o,\ \varphi \in \underline{\underline{L}};$$

$$|\int\alpha(xy^{-1})f_{kl}(z:\varphi:y)dy| \leq \int|\alpha(xy^{-1})|c_2\|\varphi\|_{\underline{\underline{L}}}\,\|y\|^M dy =$$

$$= c_2\|\varphi\|_{\underline{\underline{L}}}\int|\alpha(y^{-1})|\,\|yx\|^M dy \leq c_3\|\varphi\|_{\underline{\underline{L}}}\,\|x\|^M$$ and the same for

$\int\alpha(xy^{-1})f_{kl}(-z:\underline{\underline{c}}(z)\varphi:y)dy$. Hence there are c and N such that

$$|(\alpha * E(z:\varphi))(x)| \leq c\|x\|^N\,\|\varphi\|_{\underline{\underline{L}}} \quad \text{for } x \in G,\ \varphi \in \underline{\underline{L}},\ z \in D_o.$$

So $|E(z:\varphi:x)| \leq c\|x\|^N\,\|\pi(z:\alpha)^{-1}\varphi\|_{\underline{\underline{L}}} \leq c'\|x\|^N\,\|\varphi\|_{\underline{\underline{L}}} \quad (z \in D_o)$.

We resume our results:

Lemma 93. For any $x \in G$, $\varphi \in \underline{\underline{L}}$, the function $z \longmapsto E(z:\varphi:x)$ extends to a holomorphic function on D. As a function of (z,x), $E(z:\varphi:x)$ is C^∞ on $D \times G$. Given any compact subset D_o of D, there are constants c and N such that

$$|E(z:\varphi:x)| \leq c\|x\|^N\,\|\varphi\|_{\underline{\underline{L}}}$$

for all $\varphi \in \underline{\underline{L}}$, $z \in D_o$, $x \in G$.

It follows now immediately that $E(z:\varphi) \in L(z)$ for $z \in D$, more precisely: $E(z:\varphi) = f(h:z)$ with $h = (\varphi,\underline{\underline{c}}(z)\varphi) \in \underline{\underline{\Phi}}(z)$.

Since $\varphi \longmapsto E(z:\varphi)$ is an injective map $\underline{\underline{L}} \to L(z)$, $\dim \underline{\underline{\Phi}}(z) = \dim L(z) \geq \dim \underline{\underline{L}}$. Together with Lemma 85a) this implies

Lemma 94. $\dim \underline{\underline{\Phi}}(z) = \dim \underline{\underline{L}} \quad (z \in D)$.

§6. <u>Analytic continuation to $C - \{o\}$.</u>

Let D be, as in §5, the complement of $[-|Q|,O[$ in the half-plane $\mathrm{Re}(z) < o$ and let D' be the set $\{z \in C : \mathrm{Re}(z) \leq o,\ z \neq o\}$.

<u>Lemma 95</u>. $\dim \underline{\underline{\Phi}}(z) = \dim \underline{\underline{L}}$ for $z \in D'$ and $\underline{\underline{\Phi}}(z)$ varies continuously with z for $z \in D'$.

<u>Proof</u>. For $z \in D$ the mapping $\varphi \longmapsto (\varphi, \underline{\underline{c}}(z)\varphi)$ is a bijection of $\underline{\underline{L}}$ onto $\underline{\underline{\Phi}}(z)$ (Lemma 94), so $\underline{\underline{\Phi}}(z)$ depends continuously on z for $z \in D$. Fix $z \in D'$. If (z_n) is a sequence of points of D which converges to z, then after taking a subsequence we may assume that $\lim \underline{\underline{\Phi}}(z_n)$ exists, and then $\lim \underline{\underline{\Phi}}(z_n) \subset \underline{\underline{\Phi}}(z)$ by Lemma 60; so $\dim \underline{\underline{\Phi}}(z) \geq \dim \underline{\underline{L}}$. Assume $z \notin D$. Then $\bar{z} = \pm z$ and, by Lemma 85a), $\dim \underline{\underline{\Phi}}(z) \leq \dim \underline{\underline{L}}$. So $\dim \underline{\underline{\Phi}}(z) = \dim \underline{\underline{L}}$ for $z \in D'$; this fact and Lemma 60 give the continuity.

<u>Lemma 96</u>. If $z_o \in iR$, $z_o \neq O$, then $\lim\limits_{\substack{z \in D \\ z \to z_o}} \underline{\underline{c}}(z)$ exists.

<u>Proof</u>. Let (z_n) be a sequence of points of D which converges to z_o. Consider, for $\varphi \in \underline{\underline{L}}$, the sequence $(\underline{\underline{c}}(z_n)\varphi)$. This sequence is bounded. For, otherwise suppose that $\|\underline{\underline{c}}(z_n)\varphi\|_{\underline{\underline{L}}}$ tends to infinity, put $\varphi_n = \|\underline{\underline{c}}(z_n)\varphi\|_{\underline{\underline{L}}}^{-1}\varphi$; then $\lim \varphi_n = o$ and $\|\underline{\underline{c}}(z_n)\varphi_n\| = 1$. The sequence $(\underline{\underline{c}}(z_n)\varphi_n)$ has a convergent subsequence, suppose that $\lim \underline{\underline{c}}(z_n)\varphi_n = \psi$. Now $\lim (\varphi_n, \underline{\underline{c}}(z_n)\varphi_n) = (O,\psi)$ in $\underline{\underline{\Phi}}$ and $(\varphi_n, \underline{\underline{c}}(z_n)\varphi_n) \in \underline{\underline{\Phi}}(z_n)$, hence $(O,\psi) \in \underline{\underline{\Phi}}(z_o)$ (Lemma 60). Since $\bar{z}_o = -z_o$, Lemma 85b) gives now $\psi = o$. This gives a contradiction, so the sequence is bounded. Suppose there are two subsequences $(\underline{\underline{c}}(z_{i_n})\varphi)$ and $(\underline{\underline{c}}(z_{j_n})\varphi)$ which converge to ψ and ψ' respectively. Then, since $(\varphi, \underline{\underline{c}}(z_n)\varphi) \in \underline{\underline{\Phi}}(z_n)$, we have $(\varphi,\psi) \in \underline{\underline{\Phi}}(z_o)$ and $(\varphi,\psi') \in \underline{\underline{\Phi}}(z_o)$ (Lemma 60), hence $(O,\psi-\psi') \in \underline{\underline{\Phi}}(z_o)$ and $\psi = \psi'$ (Lemma 85b)). So the sequence $(\underline{\underline{c}}(z_n)\varphi)$ is convergent. This proves the lemma.

<u>Definition</u>. $\underline{\underline{c}}(iy) = \lim\limits_{\substack{z \in D \\ z \to iy}} \underline{\underline{c}}(z) \qquad (y \in R,\ y \neq o).$

Again by Lemma 60, we have $(\varphi, \underline{\underline{c}}(iy)\varphi) \in \underline{\underline{\Phi}}(iy)$, hence, by Lemma 85b), $\|\underline{\underline{c}}(iy)\varphi\| = \|\varphi\|$. If $z \in D$, $z \to iy$, then $\underline{\underline{c}}(z)^* \to \underline{\underline{c}}(iy)^*$, but also $\underline{\underline{c}}(z)^* = \underline{\underline{c}}(\bar{z}) \to \underline{\underline{c}}(-iy)$, so

$\underline{\underline{c}}(iy)^* = \underline{\underline{c}}(-iy)$. These two observations give

$$\underline{\underline{c}}(iy)\ \underline{\underline{c}}(-iy) = 1 \qquad (y \in R,\ y \neq o).$$

Lemma 97. Fix $z_o \in R$ with $-|\varrho| \leq z_o < o$. Denote by π_1, π_2 the projections of $\underline{\underline{\Phi}} = \underline{\underline{L}} \times \underline{\underline{L}}$ on the first resp. second factor. There is a finite set F of complex numbers such that, if $\mu \notin F$, then $\pi_1 + \mu\pi_2$ defines a bijection of $\underline{\underline{\Phi}}(z_o)$ on $\underline{\underline{L}}$.

Proof. Let W denote the kernel of π_1 in $\underline{\underline{\Phi}}(z_o)$ and W' the orthogonal complement of W in $\underline{\underline{\Phi}}(z_o)$; π_2 is injective on W and π_1 is injective on W'. It follows from the Maass-Selberg relation that $\pi_1 W'$ and $\pi_2 W$ are orthogonal in $\underline{\underline{L}}$ (see the proof of Lemma 85 in §5). Moreover $\dim \pi_1 W' + \dim \pi_2 W = \dim W' + \dim W = \dim(W+W') = \dim \underline{\underline{\Phi}}(z_o) = \dim \underline{\underline{L}}$ (Lemma 95). So $\pi_1 W' + \pi_2 W = \underline{\underline{L}}$.

If $\psi \in \pi_2 W$, then $(o,\psi) \in W$; if $\varphi \in \pi_1 W'$, then $(\varphi, T\varphi) \in W'$, where T is a linear mapping of $\pi_1 W'$ into $\underline{\underline{L}}$; write $T\varphi = T_1\varphi + T_2\varphi$ with $T_1\varphi \in \pi_1 W'$, $T_2\varphi \in \pi_2 W$. The map $\pi_1 + \mu\pi_2$ can be represented by the matrix

$$\begin{pmatrix} \mu & \mu T_2 \\ o & 1+\mu T_1 \end{pmatrix},$$

$\pi_1 + \mu\pi_2$ is bijective if $\mu \det(1+\mu T_1) \neq 0$.

Now we consider $\underline{\underline{c}}(z)$ in the neighbourhood of z_o, $z_o \in R$, $-|\varrho| \leq z_o < o$. Fix $\mu \in C$ such that $\pi_1 + \mu\pi_2$ defines a bijection of $\underline{\underline{\Phi}}(z_o)$ on $\underline{\underline{L}}$. Since $\underline{\underline{\Phi}}(z)$ depends continuously on z, $\pi_1 + \mu\pi_2$ is injective on $\underline{\underline{\Phi}}(z)$ if z is close to z_o. So we can choose $\varepsilon > o$ and $a > o$ with the following properties: 1°. $\mathrm{Re}(z) < o$ if $|z-z_o| \leq \varepsilon$, 2°. $\|(\pi_1 + \mu\pi_2)h\|_{\underline{\underline{L}}} \geq a\|h\|_{\underline{\underline{\Phi}}}$ if $|z-z_o| \leq \varepsilon$ and $h \in \underline{\underline{\Phi}}(z)$. Denote by D_ε the set $\{z : |z-z_o| < \varepsilon\}$.

If $z \in D_\varepsilon \cap D$, then $(\varphi, \underline{\underline{c}}(z)\varphi) \in \underline{\underline{\Phi}}(z)$, and $(\pi_1 + \mu\pi_2)(\varphi, \underline{\underline{c}}(z)\varphi) = (1+\mu\underline{\underline{c}}(z))\varphi$; so $\det(1+\mu\underline{\underline{c}}(z)) \neq o$ for $z \in D \cap D_\varepsilon$. Put

$$\underline{\underline{c}}_o(z) = \frac{\underline{\underline{c}}(z)}{1+\mu\underline{\underline{c}}(z)} .$$

We shall prove that $\underline{\underline{c}}_o(z)$ extends to a holomorphic function on D_ε. It is of course enough to prove that $\underline{\underline{c}}_o(z)$ extends to a continuous function on D_ε. Let (z_n) be a sequence of points of $D \cap D_\varepsilon$ which converges to z^o, $z^o \in D_\varepsilon$, and let $\varphi \in \underline{\underline{L}}$. Put $\varphi_n = (1+\mu\underline{\underline{c}}(z_n))^{-1}\varphi$, $h_n = (\varphi_n, \underline{\underline{c}}(z_n)\varphi_n)$. Then $h_n \in \underline{\underline{\Phi}}(z_n)$ and $(\pi_1+\mu\pi_2)h_n = \varphi$. The sequence (h_n) is bounded, because $a\|h_n\|_{\underline{\underline{\Phi}}} \leq \|(\pi_1+\mu\pi_2)h_n\|_{\underline{\underline{L}}} = \|\varphi\|_{\underline{\underline{L}}}$. If a subsequence of (h_n) converges to h in $\underline{\underline{\Phi}}$, then $h \in \underline{\underline{\Phi}}(z^o)$ and $(\pi_1+\mu\pi_2)h = \varphi$; since $\pi_1+\mu\pi_2$ is injective on $\underline{\underline{\Phi}}(z^o)$, this implies that (h_n) converges. So $\underline{\underline{c}}_o(z_n)\varphi = \pi_2 h_n$ converges.

So $\underline{\underline{c}}_o(z)$ is holomorphic at z_o, and $\underline{\underline{c}}(z) = \dfrac{\underline{\underline{c}}_o(z)}{1-\mu\underline{\underline{c}}_o(z)}$ is meromorphic at z_o, because $\det(1-\mu\underline{\underline{c}}_o(z))$ is not identically zero $(1-\mu\underline{\underline{c}}_o(z) = (1+\mu\underline{\underline{c}}(z))^{-1})$.

From the proof of Lemma 84 we use the formula

$$((\check{E}_{h'}, (z^2-\omega_o)^{-1}\check{E}_h))_{G/\Gamma} = g(z) - \frac{e^{2z^2}}{2z}\{(\psi,\varphi)_{\underline{\underline{L}}} + (\psi,\underline{\underline{c}}(z)\varphi)_{\underline{\underline{L}}}\}$$

where g is an entire function. Now $|(z^2-\omega_o)^{-1}| \leq |\mathrm{Im}(z^2)|^{-1} = |2z_o y|^{-1}$ if $z = z_o + iy$, z_o as above, $y \in \mathbb{R}$. So the left hand side of the above equality is $\mathcal{O}(|y|^{-1})$ for $y \to 0$, hence also $(\psi,\underline{\underline{c}}(z)\varphi)_{\underline{\underline{L}}} = \mathcal{O}(|y|^{-1})$, so that $(\psi,\underline{\underline{c}}(z)\varphi)_{\underline{\underline{L}}}$ has at most a simple pole at z_o.

Obviously $\det \underline{\underline{c}}(z)$ is not identically zero, so $\underline{\underline{c}}(z)^{-1}$ is defined as a meromorphic function on the half-plane $\mathrm{Re}(z) < 0$.

<u>Definition</u>. $\underline{\underline{c}}(z) = \underline{\underline{c}}(-z)^{-1}$ when $\mathrm{Re}(z) > 0$.

If $z_n \to iy$, $\mathrm{Re}(z_n) > 0$, then $-z_n \to -iy$, $\underline{\underline{c}}(-z_n) \to \underline{\underline{c}}(-iy) = \underline{\underline{c}}(iy)^{-1}$ and $\underline{\underline{c}}(z_n) = \underline{\underline{c}}(-z_n)^{-1} \to \underline{\underline{c}}(iy)$. So $\underline{\underline{c}}(z)$ is continuous in the points iy, $y \in \mathbb{R}$, $y \neq 0$, and therefore holomorphic at those points.

We resume what we have proved:

<u>Lemma 98</u>. The function $\underline{\underline{c}}$ is meromorphic on $\mathbb{C} - \{0\}$; its poles in the half-plane $\mathrm{Re}(z) < 0$ lie in the set $\{z \in \mathbb{R}, -|Q| \leq z < 0\}$ and are simple; $\underline{\underline{c}}$ is holomorphic at the points iy, $y \in \mathbb{R}$, $y \neq 0$. Moreover $\underline{\underline{c}}(z)\underline{\underline{c}}(-z) = 1$.

Consider now $E(z:\varphi)$. For $z \in D$, this is an element of $L(z)$ (see §5); more precisely, $E(z:\varphi) = f(h:z)$ with $h = (\varphi, \underline{\underline{c}}(z)\varphi) \in \underline{\Phi}(z)$.

Let z_n tend to iy, $y \in R$, $y \neq 0$, $z_n \in D$, and put $h_n = (\varphi, \underline{\underline{c}}(z_n)\varphi)$. Then $\lim h_n = (\varphi, \underline{\underline{c}}(iy)\varphi) = h$ and $\lim f(h_n:z_n) = f(h:iy)$ uniformly on every compact subset of G (Lemma 60). So we define

$$E(iy:\varphi) = \lim_{\substack{z \in D \\ z \to iy}} E(z:\varphi).$$

Since $E(iy:\varphi) = f(h:iy) \in L(iy)$, we have

$$E(iy:\varphi)_{P_{kl}} = \varphi_{kl}\, e^{(iy-|\varrho|)t_{kl}} + (\underline{\underline{c}}(iy)\varphi)_{kl}\, e^{-(iy+|\varrho|)t_{kl}} \text{, and also}$$

$$E(-iy:\varphi')_{P_{kl}} = (\underline{\underline{c}}(-iy)\varphi')_{kl}\, e^{(iy-|\varrho|)t_{kl}} + \varphi'_{kl}\, e^{-(iy+|\varrho|)t_{kl}} .$$

Hence

$$E(iy:\varphi) = E(-iy:\underline{\underline{c}}(iy)\varphi).$$

Fix $z_o \in R$ with $-|\varrho| \leq z_o < 0$. If $\underline{\underline{c}}$ is holomorphic at z_o, then, as z tends to z_o $(z \in D)$, $(\varphi, \underline{\underline{c}}(z)\varphi)$ tends to $(\varphi, \underline{\underline{c}}(z_o)\varphi)$ and $E(z:\varphi)$ tends, uniformly on every compact subset of G, to an element of $L(z_o)$ which we call $E(z_o:\varphi)$. $E(z:\varphi:x)$ is then holomorphic at z_o. If $\underline{\underline{c}}$ has a pole at z_o, then $(z-z_o)(\varphi, \underline{\underline{c}}(z)\varphi)$ tends to $(0, A\varphi)$, where A is a linear transformation of $\underline{\underline{L}}$, and $(z-z_o)E(z:\varphi)$ tends uniformly on every compact subset of G to an element $e(\varphi)$ of $L(z_o)$; $e(\varphi)_{P_{kl}} = (A\varphi)_{kl}\, e^{-(z_o+|\varrho|)t_{kl}}$. $E(z:\varphi:x)$ is then holomorphic or has a simple pole at z_o.

We define

$$E(z:\varphi) = E(-z:\underline{\underline{c}}(z)\varphi) \quad \text{when} \quad Re(z) > 0.$$

If z tends to iy, $Re(z) > 0$, then $E(-z:\varphi)$ tends to $E(-iy:\varphi)$, by definition; hence $E(z:\varphi) = E(-z:\underline{\underline{c}}(z)\varphi)$ tends to $E(-iy:\underline{\underline{c}}(iy)\varphi) = E(iy:\varphi)$. So $E(z:\varphi:x)$ is continuous and therefore holomorphic at iy.

If $\underline{\underline{c}}$ is holomorphic at z_o, $Re(z_o) > 0$, then $E(z:\varphi) = E(-z:\underline{\underline{c}}(z)\varphi) = f(h'(z):-z) = f(h(z):z)$ with $h'(z) = (\underline{\underline{c}}(z)\varphi, \varphi)$, $h(z) = (\varphi, \underline{\underline{c}}(z)\varphi)$ for z in a neighbourhood of z_o, $z \neq z_o$. By Corollary 2 of Lemma 60, $E(z:\varphi:x)$ is holomorphic at z_o.

We resume what we have proved:

Lemma 99. For any $x \in G$, $\varphi \in \underline{\underline{L}}$, the function $z \longmapsto E(z:\varphi:x)$ is meromorphic on $\mathbb{C} - \{o\}$; its poles are poles of $\underline{\underline{c}}$; if it has a pole at z_o with $z_o \in R$, $-|\varrho| \leq z_o < 0$, then this pole is simple. Moreover we have

$$E(z:\varphi) = E(-z:\underline{\underline{c}}(z)\varphi).$$

§7. Analytic continuation: the point 0.

Consider again the spaces $L(z)$ and $\underline{\underline{\Phi}}(z)$ defined in §5. In their definition occur representations $\chi_{kl}(z)$ of A_{kl} (see Chap.III,§1; they have not been written down explicitly in §5 of this chapter). On the base $(\frac{1}{2}(1 \pm z^{-1}H_{kl}))$ of B_{kl}, $\chi_{kl}(z:\exp tH_{kl})$ is represented by the matrix

$$\begin{pmatrix} e^{(z-|\varrho|)t} & 0 \\ 0 & e^{-(z+|\varrho|)t} \end{pmatrix},$$

so on the base $(1,H_{kl})$ it is represented by the matrix

$$e^{-|\varrho|t}\begin{pmatrix} \cosh zt & z \sinh zt \\ z^{-1}\sinh zt & \cosh zt \end{pmatrix}.$$

Using the bases $(1,H_{kl})$ we get a new identification of $\Pi\, L_{kl} \otimes B^*_{kl}$ with $\underline{\underline{L}} \times \underline{\underline{L}}$, and from this we get: $L(z)$ is the space of all functions f in $\mathcal{A}_1(G/\Gamma,\sigma)$ for which there exists an element $h = (\varphi,\psi) \in \underline{\underline{L}} \times \underline{\underline{L}} = \underline{\underline{\Phi}}$ such that

1) $f_P = 0$ if $P \in \mathcal{P}_1^o$, $P \notin \mathcal{C}^o$,

2) $f_{P_{kl}}(x) = e^{-|\varrho|t_{kl}(x)}\{\varphi_{kl}(x)\cosh zt_{kl}(x) + \psi_{kl}(x)\dfrac{\sinh zt_{kl}(x)}{z}\}$

This new definition for the same space $L(z)$ is also valid for $z = 0$. We define the space $\underline{\underline{\Phi}}(z)$ and the bijective mapping $h \longmapsto f(h:z)$ of $\underline{\underline{\Phi}}(z)$ on $L(z)$ according to Chap.III,§1; this is not the same $\underline{\underline{\Phi}}(z)$ as in §§5 and 6 [if (φ,ψ) belongs to the old $\underline{\underline{\Phi}}(z)$, then $(\varphi+\psi, z(\varphi-\psi))$ belongs to the new $\underline{\underline{\Phi}}(z)$]. We have $E(z:\varphi) = f(h:z)$ with $h = ((1+\underline{\underline{c}}(z))\varphi, z(1-\underline{\underline{c}}(z))\varphi)$.

One sees from Lemma 95 that $\dim \underline{\underline{\Phi}}(z) = \dim L(z) = \dim \underline{\underline{L}}$ for $z \neq 0$ and that $\underline{\underline{\Phi}}(z)$ varies continuously with z for $z \neq 0$. If $z_n \to 0$, $z_n \neq 0$, then after taking a subsequence, $\lim \underline{\underline{\Phi}}(z_n) \subset \underline{\underline{\Phi}}(0)$ by Lemma 60, so $\dim \underline{\underline{\Phi}}(0) \geq \dim \underline{\underline{L}}$. On the other hand, the Maass-Selberg relation gives

$$\sum_{k,l} J_{P_{kl}}(f,g:t') = 0 \quad \text{if } f,g \in L(0) .$$

If $f = f(h:0)$, $g = f(h':0)$ with $h,h' \in \underline{\underline{\Phi}}(0)$, $h = (\varphi,\psi)$, $h' = (\varphi',\psi')$, then $J_{P_{kl}}(f,g:t') = (\psi,\varphi')_{kl} - (\varphi,\psi')_{kl}$, so we have

$$(\psi,\varphi')_{\underline{\underline{L}}} = (\varphi,\psi')_{\underline{\underline{L}}} .$$

This gives $\dim \underline{\underline{\Phi}}(0) \leq \dim \underline{\underline{L}}$ (cf. the proof of Lemma 85a)). So $\dim \underline{\underline{\Phi}}(0) = \dim \underline{\underline{L}}$ and $\underline{\underline{\Phi}}(z)$ is a continuous function of z for all $z \in \mathbb{C}$.

Let π_1, π_2 be the projections of $\underline{\underline{\Phi}} = \underline{\underline{L}} \times \underline{\underline{L}}$ on the first, resp. second factor. We can choose $\mu \in \mathbb{C}$, $\mu \neq 0$, such that $\pi_1+\mu\pi_2$ defines a bijection of $\underline{\underline{\Phi}}(0)$ on $\underline{\underline{L}}$ (same proof as Lemma 97). Choose $\varepsilon > 0$, $a > 0$ such that $\|(\pi_1+\mu\pi_2)h\|_{\underline{\underline{L}}} \geq a\|h\|_{\underline{\underline{\Phi}}}$ for $h \in \underline{\underline{\Phi}}(z)$, $|z| \leq \varepsilon$. Let D denote the set of all points $z \neq 0$ where $\underline{\underline{c}}$ is holomorphic, and D_ε the set of all $z \in D$ with $|z| < \varepsilon$. If $z \in D$, then $((1+\underline{\underline{c}}(z))\varphi, z(1-\underline{\underline{c}}(z))\varphi) \in \underline{\underline{\Phi}}(z)$, hence

$$\det(1 + \mu z + (1-\mu z)\underline{\underline{c}}(z)) \neq 0 \quad \text{for } z \in D_\varepsilon .$$

Put

$$\underline{\underline{c}}_0(z) = (1 + \mu z + (1-\mu z)\underline{\underline{c}}(z))^{-1}(1+\underline{\underline{c}}(z)).$$

We claim that $\underline{\underline{c}}_0$ is holomorphic at $z = 0$. Let $z_n \in D_\varepsilon$, $z_n \to 0$, and fix $\varphi \in \underline{\underline{L}}$. Put $\varphi_n = (1 + \mu z_n + (1-\mu z_n)\underline{\underline{c}}(z_n))^{-1}\varphi$ and $h_n = ((1 + \underline{\underline{c}}(z_n))\varphi_n, z_n(1 - \underline{\underline{c}}(z_n))\varphi_n)$. Then $(\pi_1 + \mu\pi_2)h_n = \varphi$ and $h_n \in \underline{\underline{\Phi}}(z_n)$. The sequence (h_n) is bounded, hence converget (because $\pi_1 + \mu\pi_2$ is injective on $\underline{\underline{\Phi}}(0)$). So $\lim(1+\underline{\underline{c}}(z_n))\varphi_n$ exists, i.e. $\lim \underline{\underline{c}}_0(z_n)\varphi$ exists. So $\underline{\underline{c}}_0$ is holomorphic at $z = 0$.
Hence $\underline{\underline{c}}(z) = (1-(1-\mu z)\underline{\underline{c}}_0(z))^{-1}((1+\mu z)\underline{\underline{c}}_0(z)-1)$ is meromorphic at $z = 0$, because $\det(1-(1-\mu z)\underline{\underline{c}}_0(z))$ is not identically zero $(1-(1-\mu z)\underline{\underline{c}}_0(z) = 2\mu z(1+\mu z+(1-\mu z)\underline{\underline{c}}(z))^{-1})$.

Since $\underline{\underline{c}}(iy)$ is unitary, we conclude that $\underline{\underline{c}}$ is holomorphic at 0 .

Lemma 100. $\underline{\underline{c}}$ is holomorphic at 0 ; $E(z:\varphi:x)$ is holomorphic at 0.

Proof. The first assertion has been proved and the second one follows from the first one: if $z \to o$, then $((1+\underline{\underline{c}}(z))\varphi , z(1-\underline{\underline{c}}(z))\varphi) \to ((1+\underline{\underline{c}}(0))\varphi,0)$ and $E(z:\varphi)$ tends to an element $E(o:\varphi)$ of $L(o)$ (Lemma 60).

We resume the results of §§5-7 in

Theorem 7. 1°. $\underline{\underline{c}}$ is meromorphic on C ; it has a finite number of simple poles in the real interval $[-|\varrho|,0[$ and the other poles lie in $\{z:Re(z) > 0\}$.

2°. $\underline{\underline{c}}(z)\underline{\underline{c}}(-z) = 1$.

3°. For any $x \in G$, $\varphi \in \underline{\underline{L}}$, $E(z:\varphi:x)$ is meromorphic on C ; its poles are poles of $\underline{\underline{c}}$; its poles in $[-|\varrho|,0[$ are simple.

4°. $E(z:\varphi) = E(-z:\underline{\underline{c}}(z)\varphi)$.

5°. Let D denote the set of all points of C where $\underline{\underline{c}}$ is holomorphic and let ω be a compact subset of D . Then for any $\varphi \in \underline{\underline{L}}$, $E(z:\varphi:x)$ is continuous on $D \times G$ and there exist numbers c and N such that

$$|E(z:\varphi:x)| \leq c\|\varphi\|_{\underline{\underline{L}}} \|x\|^N \qquad \text{for } z \in \omega,\ \varphi \in \underline{\underline{L}},\ x \in G .$$

(The assertions under 5° follow from Theorem 6 (together with Lemma 54) and Lemma 60).

§8. Boundedness of $\underline{\underline{c}}(z)$ for $Im(z) \to \infty$.

In order to find the spectral decomposition of $L^2(G/\Gamma)$ one necessary thing to do is to shift the integration in the formula for the scalar product $((\check{E}_h,\check{E}_h))_{G/\Gamma}$ (see §5) to the imaginary axis. We prove in this paragraph a lemma which is to be used to estimate the integrals over segments joining the lines $Re(z) = c$ and $Re(z) = 0$. The notations are as in §5.

Lemma 101. Fix $u_o < o$, $y_o > 0$. $|\underline{\underline{c}}(z)|$ remains bounded for $z = u + iy$, $u_o \leq u \leq 0$, $|y| \geq y_o$.

Proof. We call the number t which occurs in expressions like $\tilde{S}'_{kl}(t)$ here $-N$. Define a function v on $\mathbb{R}$ by $v(x) = 0$ for $x \leq -N$, $v(x) = 1$ for $x > -N$, and define functions v_{kl} on G by $v_{kl}(x) = v(t_{kl}(x))$. Define for $Re(z) < -|\varrho|$, $\varphi \in \underline{L}$:

$$E^{(1)}(z:\varphi:x) = \sum_{k,l} \sum_{\Gamma/\Gamma\cap P_{kl}} v_{kl}(x\gamma)\varphi_{kl}(x\gamma)e^{(z-|\varrho|)t_{kl}(x\gamma)} ,$$

$$E^{(2)}(z:\varphi:x) = \sum_{k,l} \sum_{\Gamma/\Gamma\cap P_{kl}} (1-v_{kl}(x\gamma))(\underline{c}(z)\varphi)_{kl}(x\gamma)e^{-(z+|\varrho|)t_{kl}(x\gamma)} ,$$

$$E^{(3)}(z:\varphi:x) = E^{(1)}(z:\varphi:x) - E^{(2)}(z:\varphi:x).$$

Fix z_o and R such that $R > |\varrho|$, $Re(z_o) < -R$, and put $v_o(t) = v(t)e^{(z_o-|\varrho|)t}$ and $v'_o(t) = (1-v(t))e^{-(z_o+|\varrho|)t}$. Then the integrals

$\int_{-\infty}^{\infty} |v_o(t)|^2 e^{2|\varrho|t}(e^{2Rt} + e^{-2Rt})dt$ and $\int_{-\infty}^{\infty} |v'_o(t)|^2 e^{2|\varrho|t}(e^{2Rt} + e^{-2Rt})dt$ are finite;

we compute the Fourier transforms $\hat{v}_o$ and $\hat{v}'_o$:

$$\hat{v}_o(z) = \int_{-\infty}^{\infty} v_o(t)\, e^{-(z-|\varrho|)t}\, dt = \frac{e^{(z-z_o)N}}{z-z_o} \quad \text{and}$$

$$\hat{v}'_o(z) = \int_{-\infty}^{\infty} v'_o(t)\, e^{-(z-|\varrho|)t}\, dt = -\frac{e^{(z+z_o)N}}{z+z_o} .$$

This shows that, if we put

$$h(z) = \frac{e^{(z-z_o)N}}{z-z_o}\varphi + \frac{e^{(z+z_o)N}}{z+z_o}\underline{c}(z_o)\varphi ,$$

then $h \in \mathcal{H}(R)$, so that $\check{E}_h$ is defined (§4), and obviously $\check{E}_h = E^{(3)}(z_o:\varphi)$.

Let now $z_1,z_2 \in \mathbb{C}$ with $Re(z_i) < -R < -|\varrho|$ and let $\varphi,\psi \in \underline{L}$. Put

$$h_1(z) = \frac{e^{(z-z_1)N}}{z-z_1}\varphi + \frac{e^{(z+z_1)N}}{z+z_1}\underline{c}(z_1)\varphi ,$$

$$h_2(z) = \frac{e^{(z-z_2)N}}{z-z_2}\psi + \frac{e^{(z+z_2)N}}{z+z_2}\underline{c}(z_2)\psi .$$

Then $E^{(3)}(z_1:\varphi) = \check{E}_{h_1}$ and $E^{(3)}(z_2:\psi) = \check{E}_{h_2}$, so we have (see §5):

$$((E^{(3)}(z_2:\psi), E^{(3)}(z_1:\varphi)))_{G/\Gamma} = \frac{1}{2\pi i}\int_{c-i\infty}^{c+i\infty}\{(h_2(-\bar{z}), h_1(z))_{\underline{\underline{L}}} + (h_2(\bar{z}), \underline{\underline{c}}(z)h_1(z))_{\underline{\underline{L}}}\}dz$$

where $-R < c < -|\varrho|$. We shift the integration to a line $\mathrm{Re}(z) = c_1$ and make c_1 tend to $-\infty$. One sees from Lemma 38 that there are constants a, b such that

$$|\underline{\underline{c}}(z)| \leq \frac{a\, e^{-bc_1}}{|c_1 + |\varrho||} \qquad \text{for } z = c_1 + iy .$$

This implies that the integral over the line $\mathrm{Re}(z) = c_1$ tends to zero as $c_1 \to -\infty$, so only the residus of the integrand at z_1 and $\bar{z}_2$ are left:

$$((E^{(3)}(z_2:\psi), E^{(3)}(z_1:\varphi)))_{G/\Gamma} = -\frac{e^{-(z_1+\bar{z}_2)N}}{z_1+\bar{z}_2}(\psi,\varphi) +$$

$$+ \frac{e^{(z_1-\bar{z}_2)N}(\psi,\underline{\underline{c}}(z_1)\varphi) - e^{-(z_1-\bar{z}_2)N}(\psi,\underline{\underline{c}}(\bar{z}_2)\varphi)}{z_1-\bar{z}_2} + \frac{e^{(z_1+\bar{z}_2)N}}{z_1+\bar{z}_2}(\psi,\underline{\underline{c}}(\bar{z}_2)\underline{\underline{c}}(z_1)\varphi) .$$

Call the right hand side of this formula $\Omega_{\psi,\varphi}(\bar{z}_2:z_1)$. Then $\Omega_{\psi,\varphi}$ is holomorphic on $D \times D$ where D is, as in §5, the complement of the interval $]-|\varrho|,0[$ in the half-plane $\mathrm{Re}(z) < 0$. One proves now in exactly the same way as Lemma 87 was proved that the mapping $z \longmapsto E^{(3)}(z:\varphi)$ extends to a holomorphic mapping $D \to L^2(G/\Gamma,\sigma)$.

If $z = u + iy \in D$, then

$$\|E^{(3)}(z:\varphi)\|^2_{G/\Gamma} = \frac{e^{2uN}|\underline{\underline{c}}(z)\varphi|^2 - e^{-2uN}|\varphi|^2}{2u} + \mathrm{Re}\,\frac{e^{2iyN}(\varphi,\underline{\underline{c}}(z)\varphi)}{iy} ,$$

hence

$$-\frac{e^{2uN}|\underline{\underline{c}}(z)\varphi|^2 - e^{-2uN}}{2|u|} + \frac{|\underline{\underline{c}}(z)\varphi|}{|y|} \geq 0 \quad \text{if } |\varphi| = 1 .$$

This gives $|\underline{\underline{c}}(z)\varphi| \leq e^{-2uN}\left(|\frac{u}{y}| + \sqrt{1+\frac{u^2}{y^2}}\right)$, so $|\underline{\underline{c}}(z)| \leq e^{-2uN}(1 + 2|\frac{u}{y}|)$.

Lemma 101 is proved.

The result of shifting the integration in the formula for $((\check{E}_{h'},\check{E}_h))_{G/\Gamma}$ is

$$((\check{E}_{h'},\check{E}_h))_{G/\Gamma} = - \sum_{1 \leq j \leq N} (h'(z_j), C_j h(z_j))_{\underline{\underline{L}}} +$$

$$+ \frac{1}{2\pi} \int_{-\infty}^{\infty} \{(h'(iy),h(iy))_{\underline{\underline{L}}} + (h'(-iy),\underline{\underline{c}}(iy)h(iy))_{\underline{\underline{L}}}\} \, dy ,$$

where $z_1,\ldots,z_N$ are the poles of $\underline{\underline{c}}(z)$ in $[-|\varrho|,0[$ and C_j is the residu of $\underline{\underline{c}}(z)$ at z_j. This formula is certainly valid if h and h' are the Fourier transforms of two functions in $C_c^\infty(A)$.

CHAPTER V

§1. Some preparation.

If P is a cuspidal subgroup of G and ξ an element of $\mathcal{H}^{\circ}(P)$, we can define a space $L_P(\sigma,\xi)$ by

$$L_P(\sigma,\xi) = {}^{\circ}L^2(M/\Gamma_M,\sigma_M,\xi) = \sum_{\chi\in\xi} {}^{\circ}L^2(M/\Gamma_M,\sigma_M,\chi) ;$$

regarded as a space of functions on G it is independant of the split component A of P (see Chap.IV,§4).

Let P_1, P_2 be two associated cuspidal subgroups of G. Choose split components A_1, A_2 in P_1, P_2 and let $s \in w(\mathfrak{a}_1,\mathfrak{a}_2)$ and $\Lambda \in (\mathfrak{a}^*_{1,c})^-$. By Theorem 5 we have a linear mapping $c(s:\Lambda)$ of $L_{P_1}(\sigma,\xi_1)$ into $L_{P_2}(\sigma,\xi_2)$ for $\xi_1 \in \mathcal{H}^{\circ}(P_1)$, $\xi_2 = i_{P_1|P_2}(\xi_1) \in \mathcal{H}^{\circ}(P_2)$ (for $i_{P_1|P_2}$ see Chap.IV,§4). If $A_1' = {}^{y_1}A_1$ and $A_2' = {}^{y_2}A_2$ are other split components of P_1 and P_2 $(y_i \in (P_i)_Q)$, then $c(y_2 s y_1^{-1} : {}^{y_1}\Lambda) = c(s:\Lambda)$. We shall denote this mapping by $c_{P_2|P_1}(s:\Lambda)$.

<u>Lemma 102</u>. Let P_1, P_2 be two associated cuspidal subgroups of G. Let $\gamma_i \in \Gamma$, $p_i \in (P_i)_Q$ $(i = 1,2)$ and put $P_i' = {}^{\gamma_i}P_i$. Then for any $s \in w(\mathfrak{a}_1,\mathfrak{a}_2)$ and $\Lambda \in (\mathfrak{a}^*_{1,c})^-$, where $\mathfrak{a}_i$ is a split component of $\mathfrak{P}_i$, we have

$$c_{P_2'|P_1'}(\gamma_2 p_2 s p_1^{-1}\gamma_1^{-1} : {}^{\gamma_1 p_1}\Lambda) = e^{(s\Lambda-\varrho_2)(H_2(\gamma_2))}\,\tau_{\gamma_2} c_{P_2|P_1}(s:\Lambda)\tau_{\gamma_1}^{-1}\, e^{-(\Lambda-\varrho_1)(H_1(\gamma_1))} .$$

Here τ_{γ_i} is defined by $(\tau_{\gamma_i}\varphi)(x) = \varphi(x\gamma_i)$.

<u>Proof</u>. If we denote by $E(P|A:\Lambda:\varphi)$ the Eisenstein series which was called $E(\Lambda:\varphi)$ in Chap.II,§2, we have

$$E({}^{\gamma}P_1|{}^{\gamma}A_1:{}^{\gamma}\Lambda:\tau_{\gamma}\varphi) = e^{-(\Lambda-\varrho_1)(H_1(\gamma))}\, E(P_1|A_1:\Lambda:\varphi) ,$$

which gives

$$c_{P_2|{}^{\gamma}P_1}(s\gamma^{-1}:{}^{\gamma}\Lambda)\tau_{\gamma} = e^{-(\Lambda-\varrho_1)(H_1(\gamma))}\, c_{P_2|P_1}(s:\Lambda) .$$

From $f_{\gamma_{P_2}}(x) = f_{P_2}(x\gamma)$, where one takes $f = E(P_1|A_1:\Lambda:\varphi)$, one sees that

$$c_{\gamma_{P_2|P_1}}(\gamma s:\Lambda) = e^{(s\Lambda-\varrho_2)(H_2(\gamma))}\, \tau_\gamma\, c_{P_2|P_1}(s:\Lambda).$$

We remarked already that $c_{P_2|P_1}(p_2 s p_1^{-1}: {}^{p_1}\Lambda) = c_{P_2|P_1}(s:\Lambda)$; all this together gives the formula.

In Lemmas 103-105, (P,A) is a cuspidal pair of G dominated by a cuspidal pair (P',A'), $({}^*P,{}^*A)$ denotes the cuspidal pair $(M' \cap P,\ M' \cap A)$ of M' and $\pi' = \pi_{P'|M'}$.

<u>Lemma 103</u>. $\pi'(\Gamma \cap P) = \Gamma_{M'} \cap {}^*P$ and π' defines a bijection of $\Gamma \cap P'/\Gamma \cap P$ on $\Gamma_{M'}/\Gamma_{M'} \cap {}^*P$.

<u>Proof</u>. Trivial.

<u>Lemma 104</u>. For any $\Lambda \in \mathfrak{a}_c^*$ let ${}^*\Lambda$ denote the restriction of Λ on ${}^*\mathfrak{a}_c$. Then $\Lambda \in (\mathfrak{a}_c^*)^-$ implies ${}^*\Lambda \in ({}^*\mathfrak{a}_c^*)^-$.

<u>Proof</u>. Extend $\mathfrak{a}$ to a Cartan subalgebra $\mathfrak{f}$ of $\mathfrak{g}$. For any root β of $(\mathfrak{g},\mathfrak{f})$ put $H_\beta' = 2\,H_\beta/\beta(H_\beta)$. Then, if β is a root also of $\mathfrak{m}'$, H_β' is independant of whether you regard β as a root of $\mathfrak{g}$ or of $\mathfrak{m}'$. Hence for any root α in $\Sigma(P|A)$ such that $\alpha = 1$ on $\mathfrak{a}'$ and any $\lambda \in \mathfrak{a}^*$, $\langle\lambda,\alpha\rangle$ and $\langle{}^*\lambda,{}^*\alpha\rangle$ differ by a strictly positive factor. If $\varrho = \varrho_{P|A}$, then the restriction ${}^*\varrho$ of ϱ on ${}^*\mathfrak{a}$ is just $\varrho_{{}^*P|{}^*A}$. So, if $\langle\lambda+\varrho,\alpha\rangle < 0$ for all $\alpha \in \Sigma(P|A)$, then $\langle{}^*\lambda+{}^*\varrho,{}^*\alpha\rangle < 0$ for all $\alpha \in \Sigma(P|A)$ such that $\alpha = 1$ on $\mathfrak{a}'$, i.e. if $\lambda \in (\mathfrak{a}^*)^-$ then ${}^*\lambda \in ({}^*\mathfrak{a}^*)^-$.

The Eisenstein series $E(P|A:\Lambda:\varphi)$ is defined for $\Lambda \in (\mathfrak{a}_c^*)^-$, $\varphi \in L_P(\sigma,\xi)$, $\xi \in \mathcal{X}^0(P)$. If $\varphi \in L_P(\sigma,\xi)$, $\Lambda \in (\mathfrak{a}_c^*)^-$, then $E({}^*P|{}^*A:{}^*\Lambda:\varphi)$ is defined as a function on M'; we extend it to a σ-function on $G/A'U'$.

<u>Lemma 105</u>. For $\varphi \in L_P(\sigma,\xi)$, $\Lambda \in (\mathfrak{a}_c^*)^-$ we have

$$E(P|A:\Lambda:\varphi:x) = \sum_{\Gamma/\Gamma\cap P'} E({}^*P|{}^*A:{}^*\Lambda:\varphi:x\gamma)\ e^{(\Lambda-\varrho)(H'(x\gamma))} \qquad (x \in G),$$

the series being absolutely convergent.

Proof. Put $H(x) = {}^*H(x) + H'(x)$ with ${}^*H(x) \in {}^*\mathfrak{a}$ and $H'(x) \in \mathfrak{a}'$. Then $H' = H_{P'|A'}$ and ${}^*H(m') = H_{*P|*A}(m')$ for $m' \in M'$. Now

$$E(P|A:\Lambda:\varphi:x) = \sum_{\gamma\in\Gamma/\Gamma\cap P'} \sum_{\delta\in\Gamma\cap P'/\Gamma\cap P} \varphi(x\gamma\delta)\, e^{(\Lambda-\varrho)(H'(x\gamma)+{}^*H(x\gamma\delta))},$$

$$\sum_{\Gamma\cap P'/\Gamma\cap P} \varphi(y\delta)\, e^{(\Lambda-\varrho)({}^*H(y\delta))} = \sum_{\Gamma_{M'}/\Gamma_{M'}\cap{}^*P} \varphi(y\delta)\, e^{({}^*\Lambda-{}^*\varrho)({}^*H(y\delta))} .$$

Lemma 106. Let (P_1,A_1) and (P_2,A_2) be two associated cuspidal pairs of G. Let (P',A') be a cuspidal pair and y_1, y_2 elements of $G_{\mathbb{Q}}$ such that ${}^{y_i}(P',A') \succ (P_i,A_i)$ $(i = 1,2)$. Then ${}^{\gamma\, y_1}P' = {}^{y_2}P'$ for every $\gamma \in \Gamma(s)$, $s \in w(\mathfrak{a}_1,\mathfrak{a}_2)$ provided $s = \mathrm{Ad}(y_2y_1^{-1})$ on ${}^{y_1}(\mathfrak{a}')$.

Proof. If s is induced by ξ, then $(y_2y_1^{-1})^{-1}\xi \in {}^{y_1}P'$, hence $P_2\xi P_1 \subset y_2P'y_1^{-1}$ and $\Gamma(s) \subset y_2P'y_1^{-1}$.

Corollary. $c_{P_2|P_1}(s:\Lambda) = 0$ for $s \in w(\mathfrak{a}_1,\mathfrak{a}_2)$ such that $s = \mathrm{Ad}(y_2y_1^{-1})$ on ${}^{y_1}(\mathfrak{a}')$ unless ${}^{y_1}P'$ and ${}^{y_2}P'$ are conjugate under Γ.

Proof. If ${}^{y_1}P'$ and ${}^{y_2}P'$ are not conjugate under Γ, then $\Gamma(s)$ is empty by Lemma 106.

In Lemmas 107-108 we suppose that (P_1,A_1) and (P_2,A_2) are associated cuspidal pairs of G which are both dominated by (P',A'). We put $({}^*P_i,{}^*A_i) = (M'\cap P_i,\, M'\cap A_i)$ $(i = 1,2)$ and denote by $w'(\mathfrak{a}_1,\mathfrak{a}_2)$ the set of those elements s of $w(\mathfrak{a}_1,\mathfrak{a}_2)$ which induce the identity on $\mathfrak{a}'$. For $s \in w'(\mathfrak{a}_1,\mathfrak{a}_2)$, ${}^*s$ denotes the restriction of s on ${}^*\mathfrak{a}_1$.

Lemma 107. The mapping $s \longmapsto {}^*s$ is a bijection of $w'(\mathfrak{a}_1,\mathfrak{a}_2)$ on $w({}^*\mathfrak{a}_1,{}^*\mathfrak{a}_2)$.

Proof. If an element s of $w'(\mathfrak{a}_1,\mathfrak{a}_2)$ is induced by an element ξ of $G_{\mathbb{Q}}$, then $\xi \in (M'A')_{\mathbb{Q}}$; if $\xi = \xi_o a_o$ with $\xi_o \in M'_{\mathbb{Q}}$, $a_o \in A'_{\mathbb{Q}}$, then $s = \mathrm{Ad}(\xi_o)$ on $\mathfrak{a}_1$, and the lemma is obvious.

Lemma 108. For $s \in w'(\mathfrak{a}_1,\mathfrak{a}_2)$, $\Lambda \in (\mathfrak{a}^*_{1,c})^-$, we have

$$c_{P_2|P_1}(s:\Lambda) = c_{*P_2|*P_1}(*s:*\Lambda).$$

Proof. Fix $\varphi \in {}^0L^2(M_1/\Gamma_{M_1},\sigma_{M_1},\chi)$, $\chi \in \mathcal{H}_{M_1}$. By Chap.II,§5, we have

$$(c(s:\Lambda)\varphi)(m) = \int_{U_2/U_2\cap\Gamma} \sum_{\Gamma(s)/\Gamma\cap P_1} \varphi_\Lambda(mu\gamma)\,du =$$

$$= \sum_{U_2\cap\Gamma\backslash\Gamma(s)/\Gamma\cap P_1} \int_{U_2/U_2\cap\Gamma\cap{}^\gamma P_1} \varphi_\Lambda(mu\gamma)\,du$$

for $m \in M_2$. Fix $\gamma \in \Gamma(s)$. Observe that $\gamma \in \Gamma \cap P'$. We have

$$\int_{U_2/U_2\cap\Gamma\cap{}^\gamma P_1} \varphi_\Lambda(mu\gamma)\,du = \int_{*U_2/\pi'(U_2\cap\Gamma\cap{}^\gamma P_1)} d^*u \int_{U'/U'\cap\Gamma} \varphi_\Lambda(m^*uu'\gamma)\,du'$$

$$= \int_{*U_2/\pi'(U_2\cap\Gamma\cap{}^\gamma P_1)} \varphi_\Lambda(m^*u\gamma)\,d^*u\,, \quad \text{where } \pi' = \pi_{P'|M'}.$$

Now $\pi'(U_2 \cap \Gamma \cap {}^\gamma P_1) = {}^*U_2 \cap \Gamma_{M'} \cap {}^\delta({}^*P_1)$ where $\delta = \pi'(\gamma) \in \Gamma_{M'}$. Moreover, one sees easily that $\pi'(\Gamma(s)) = \Gamma_{M'}(*s)$. So we get (see also Lemma 103):

$$(c(s:\Lambda)\varphi)(m) = \sum_{*U_2\cap\Gamma_{M'}\backslash\Gamma_{M'}(*s)/\Gamma_{M'}\cap{}^*P_1} \int_{*U_2/{}^*U_2\cap\Gamma_{M'}\cap{}^\delta({}^*P_1)} \varphi_\Lambda(m^*u\delta)\,d^*u\,.$$

Since $\varphi_\Lambda(m^*u\delta) = \varphi(m^*u\delta)\,e^{(\Lambda-\varrho_1)(H_1(m^*u\delta))} = \varphi(m^*u\delta)\,e^{(*\Lambda-{}^*\varrho_1)({}^*H_1(m^*u\delta))}$, this gives the lemma.

§2. Statement of Theorems 8 and 9.

Fix a class $\mathcal{C}$ of associated cuspidal groups. Choose some elements $P_1,\ldots,P_r$ in $\mathcal{C}$. Let $\mathcal{C}_i$ denote the class of cuspidal groups which are conjugate to P_i. Choose for each i a complete set of representatives P_{ik} $(1 \le k \le r_i)$ for $\mathcal{C}_i/\Gamma$ and fix y_{ik} in $G_{\mathbb{Q}}$ such that $P_{ik} = {}^{y_{ik}}(P_i)$.

If $\xi \in \mathcal{H}^0(\mathcal{C})$, we introduce the following notations:

$$L_{ik} = L_{P_{ik}}(\sigma,\xi_{P_{ik}}),\quad \mathbb{L}_i = \prod_{1\le k\le r_i} L_{ik}\,.$$

Let A_i be a split component of P_i $(1 \leq i \leq r)$ and put $A_{ik} = {}^{y_{ik}}(A_i)$. For $s \in w(\mathfrak{a}_i, \mathfrak{a}_j)$, $\Lambda \in (\mathfrak{a}^*_{i,C})^-$ the linear transformations

$$c_{ji}(s:\Lambda) : \mathbb{L}_i \to \mathbb{L}_j$$

are defined as in Chap.IV,§4, i.e. we have

$$(\psi, c_{ji}(s:\Lambda)\varphi)_{\mathbb{L}_j} = (\psi, c_{P_{jl}|P_{ik}}(y_{jl} s y_{ik}^{-1} : {}^{y_{ik}}\Lambda)\varphi)_{L_{jl}} \quad \text{for } \varphi \in L_{ik}, \ \psi \in L_{jl} .$$

<u>Theorem 8</u>. Fix i,j $(1 \leq i, j \leq r)$ and $s \in w(\mathfrak{a}_i, \mathfrak{a}_j)$. Then $c_{ji}(s:\Lambda)$ is a meromorphic function of Λ on $\mathfrak{a}^*_{i,C}$ all whose poles lie along hyperplanes. Moreover, if k is another index and $t \in w(\mathfrak{a}_j, \mathfrak{a}_k)$, then

$$c_{ki}(ts:\Lambda) = c_{kj}(t:s\Lambda)\, c_{ji}(s:\Lambda).$$

<u>Remarks</u>. 1. The statement of Theorem 8 is independant of the choice of the P_{ik} and the y_{ik} (see Lemma 102).
2. It is enough to prove Theorem 8 in case $\{P_1, \ldots, P_r\}$ contains a set of representatives of $\mathcal{C}/G_Q$.

Assume that $\{P_1, \ldots, P_r\}$ contains a set of representatives for $\mathcal{C}/G_Q$. Let P be any element of $\mathcal{C}$. Choose $y_k \in G_Q$ $(1 \leq k \leq p)$ such that $\{{}^{y_k}P\}$ is a set of representatives modulo Γ of the class of cuspidal groups which are conjugate to P. Let A be a split component of P and put $(Q_k, B_k) = {}^{y_k}(P,A)$. Putting

$$\mathbb{L}_P = \prod_{1 \leq k \leq p} L_{Q_k}(\sigma, \xi_{Q_k}) ,$$

where $\xi \in \mathcal{H}^o(\mathcal{C})$, we can define, for $s \in w(\mathfrak{a}, \mathfrak{a}_i)$, $\Lambda \in (\mathfrak{a}^*_C)^-$, a linear transformation

$$c_{i,P}(s:\Lambda) : \mathbb{L}_P \to \mathbb{L}_i$$

in the same way as we defined $c_{ji}(s:\Lambda)$. Let $D(P|A)$ denote the set of all $\Lambda_o \in \mathfrak{a}^*_C$ such that $c_{i,P}(s:\Lambda)$ is holomorphic at Λ_o for all i and all $s \in w(\mathfrak{a}, \mathfrak{a}_i)$. Theorem 8 implies that $D(P|A)$ is an open dense connected subset of $\mathfrak{a}^*_C$. Define for $\varphi = (\varphi_k) \in \mathbb{L}_P$, $\Lambda \in (\mathfrak{a}^*_C)^-$,

$$\mathbb{E}(P|A:\Lambda:\varphi) = \sum_{1\le k\le p} E(Q_k|B_k:{}^{y_k}\Lambda:\varphi_k) .$$

Theorem 9. Fix $\varphi \in \mathbb{L}_P$. Then $\mathbb{E}(P|A:\varphi)$ extends to a C^∞ function on $D(P|A) \times G$ such that

1) for any fixed x, $\mathbb{E}(P|A:\Lambda:\varphi:x)$ is meromorphic for $\Lambda \in \mathfrak{a}^*_C$ (and holomorphic for $\Lambda \in D(P|A)$) ;

2) given a compact set $\omega \subset D(P|A)$, we can choose numbers c and N such that

$$|\mathbb{E}(P|A:\Lambda:\varphi:x)| \le c\|\varphi\|_{\mathbb{L}_P} \|x\|^N \quad \text{for } \Lambda \in \omega,\ \varphi \in \mathbb{L}_P ,\ x \in G .$$

We shall first derive a consequence of Theorems 8 and 9 and then prove everything by induction on the rank of $\mathcal{C}$.

If $f = \mathbb{E}(P|A:\Lambda:\varphi)$, $\Lambda \in (\mathfrak{a}^*_C)^-$, $\varphi \in \mathbb{L}_P$, then, by Theorem 5:

$$f_{P_{jl}}(x) = \sum_{s\in W(\mathfrak{a},\mathfrak{a}_j)} (c_{j,P}(s:\Lambda)\varphi)_{jl}(x)\; e^{y_{jl}(s\Lambda-\varrho_j)(H_{jl}(x))} .$$

So the above theorems imply immediately:

Corollary. For $s \in W(\mathfrak{a},\mathfrak{a}_i)$, $\varphi \in \mathbb{L}_P$, we have

$$\mathbb{E}(P_i|A_i : s\Lambda : c_{i,P}(s:\Lambda)\varphi) = \mathbb{E}(P|A:\Lambda:\varphi) .$$

§3. First part of the proof of Theorem 8.

Theorems 8 and 9 are trivial if rank $\mathcal{C} = 0$ and they have been proved in Chapter IV for rank $\mathcal{C} = 1$, so we suppose now that rank $\mathcal{C} = q \ge 2$.

Let (P,A) be a cuspidal pair with $P \in \mathcal{C}$ and (P',A') a cuspidal pair with rank $P' \ge 1$ which dominates (P,A). Put $({}^*P,{}^*A) = (M' \cap P, M' \cap A)$; then $({}^*P,{}^*A)$ is a cuspidal pair of M' with rank ${}^*P < q$. We have $\mathfrak{a} = {}^*\mathfrak{a} + \mathfrak{a}'$ and write $H(x) = {}^*H(x) + H'(x)$ $(H = H_{P|A})$ as in §1. Define

$$(\mathfrak{a}^*)^+ = \{\lambda \in \mathfrak{a}^* : \langle\lambda,\alpha\rangle > 0 \quad \text{for} \quad \alpha \in \Sigma(P|A)\} ,$$

$$(\mathfrak{a}_{\mathbb{C}}^*)^+ = (\mathfrak{a}^*)^+ + i\,\mathfrak{a}^* .$$

Lemma 109. Fix a compact set $^*\omega$ in $D(^*P|^*A)$. Then we can choose $\Lambda_o' \in \mathfrak{a}'^*$ with the following property. Let Ω be a compact subset of $\Lambda_o' - \mathrm{Cl}(\mathfrak{a}_{\mathbb{C}}'^*)^+ + {}^*\omega$. Then for any compact subset C of G and any $\varphi \in L_P(\sigma,\xi_P)$ the series

$$\sum_{\Gamma/\Gamma\cap P'} |E(^*P|^*A:{}^*\Lambda:\varphi:x\gamma)\; e^{(\Lambda-\varrho)(H'(x\gamma))}|$$

converges uniformly for $\Lambda \in \Omega$ and $x \in C$. Moreover we can choose c and N such that

$$\sum_{\Gamma/\Gamma\cap P'} |E(^*P|^*A:{}^*\Lambda:\varphi:x\gamma)\; e^{(\Lambda-\varrho)(H'(x\gamma))}| \leq c\|\varphi\|\;\|x\|^N$$

for all $\Lambda \in \Omega$, $\varphi \in L_P(\sigma,\xi_P)$, $x \in G$.

Proof. By the induction hypothesis we can choose c_1, N_1 such that $|E(^*P|^*A:{}^*\Lambda:\varphi:m')| \leq c_1\|\varphi\|\;\|m'\|^{N_1}$ for $^*\Lambda \in {}^*\omega$, $\varphi \in L_P(\sigma,\xi_P)$, $m' \in M'$. Fix a mincuspidal pair $(P_o,A_o) \prec (P,A)$ and put $(^*P_o,{}^*A_o) = (M' \cap P_o,\, M' \cap A_o)$. Choose a Siegel domain $\mathfrak{S}_{M'}$ in M' with respect to the mincuspidal pair $(^*P_o,{}^*A_o)$ and a finite set $\Xi_{M'}$ in $M'_{\mathbb{Q}}$ such that $M' = \mathfrak{S}_{M'}\Xi_{M'}\Gamma_{M'}$.

$$|E(^*P|^*A:{}^*\Lambda:\varphi:m'\eta)| \leq c_2\|\varphi\|\;\|m'\|^{N_1} \leq c\|\varphi\|\; e^{(\lambda-{}^*\varrho_o)(^*H_o(m'))} \qquad \text{for all } m' \in \mathfrak{S}_{M'},$$

$\eta \in \Xi_{M'}$ with a certain $\lambda \in {}^*\mathfrak{a}_o^*$. Hence

$$|E(^*P|^*A:{}^*\Lambda:\varphi:x\eta)| \leq c\|\varphi\|\; e^{(\lambda-{}^*\varrho_o)(^*H_o(x))} \qquad \text{for } x \in K\mathfrak{S}_{M'}A'U',\ \eta \in \Xi_{M'} .$$

If $x \in K\mathfrak{S}_{M'}A'U'$, then $^*H_o(x) \in {}^*H_o - (^*\mathfrak{a}_o)^+$ with a fixed $^*H_o \in {}^*\mathfrak{a}_o$; here $(^*\mathfrak{a}_o)^+$ is the set of all points of $^*\mathfrak{a}_o$ where all positive roots of $(^*P_o,{}^*A_o)$ take strictly positive values. Choosing $N > 1$ such that $\lambda(H) \geq -N\,{}^*\varrho_o(H)$ for $H \in (^*\mathfrak{a}_o)^+$ we have $\lambda(^*H_o(x)) \leq -N\,{}^*\varrho_o(^*H_o(x)) + c'$ for $x \in K\mathfrak{S}_{M'}A'U'$, so that we may replace λ by $-N\,{}^*\varrho_o$. This shows that we may suppose that in the above inequality $\lambda \in (^*\mathfrak{a}_o^*)^-$. It is clearly possible to choose $\Lambda_o' \in \mathfrak{a}'^*$ such that $\Lambda_o' + \lambda \in (\mathfrak{a}_o^*)^-$. We claim that such a Λ_o' has the property stated in the lemma.

So let Ω be a compact subset of $\Lambda_o' - Cl(\mathfrak{a}_c'^*)^+ + {}^*\omega$ and $\Lambda \in \Omega$, $\Lambda = {}^*\Lambda + \Lambda'$ with ${}^*\Lambda \in {}^*\mathfrak{a}_C^*$, $\Lambda' \in \mathfrak{a}_C'^*$. Then ${}^*\Lambda \in {}^*\omega$ and $\Lambda' + \lambda$ stays in a compact subset of $(\mathfrak{a}_{o,C}^*)^-$. From the preceding we see that

$$|E({}^*P|{}^*A:{}^*\Lambda:\varphi:x\eta)| \leq c\|\varphi\| \sum_{{}^\eta\Gamma_{M'}/{}^\eta\Gamma_{M'}\cap{}^*P_o} e^{(\lambda-{}^*\varrho_o)({}^*H_o(x\delta))}$$

for $x \in K\mathfrak{F}_{M'}A'U'{}^\eta\Gamma_{M'}$, $\eta \in \Xi_{M'}$, so

$$|E({}^*P|{}^*A:{}^*\Lambda:\varphi:x)| \leq c\|\varphi\| \sum_{{}^\eta\Gamma_{M'}/{}^\eta\Gamma_{M'\wedge}{}^*P_o} e^{(\lambda-{}^*\varrho_o)({}^*H_o(x\eta^{-1}\delta))}$$

for $x \in K\mathfrak{F}_{M'}\eta\Gamma_{M'}A'U'$, $\eta \in \Xi_{M'}$, and

$$|E({}^*P|{}^*A:{}^*\Lambda:\varphi:x)| \leq c\|\varphi\| \sum_{\eta\in\Xi_{M'}} \sum_{\delta} e^{(\lambda-{}^*\varrho_o)({}^*H_o(x\eta^{-1}\delta))} \quad \text{for } x \in G .$$

$$|E({}^*P|{}^*A:{}^*\Lambda:\varphi:x)|\, e^{(\Lambda_R'-\varrho')(H'(x))} \leq c\|\varphi\| \sum_{\eta}\sum_{\delta} e^{(\lambda+\Lambda_R'-\varrho_o)(H_o(x\eta^{-1}\delta))} ,$$

$$\sum_{\Gamma/\Gamma\cap P'} |E({}^*P|{}^*A:{}^*\Lambda:\varphi:x\gamma)\, e^{(\Lambda-\varrho)(H'(x\gamma))}| \leq c\|\varphi\| \sum_{\gamma}\sum_{\eta}\sum_{\delta} e^{(\lambda+\Lambda_R'-\varrho_o)(H_o(x\gamma\eta^{-1}\delta))}$$

$$= c\|\varphi\| \sum_{\eta}\sum_{\gamma}\sum_{\delta} = c\|\varphi\| \sum_{\eta} \sum_{\gamma\in{}^\eta\Gamma/{}^\eta\Gamma\cap P'} \sum_{\delta} e^{(\lambda+\Lambda_R'-\varrho_o)(H_o(x\eta^{-1}\gamma\delta))}$$

$$= c\|\varphi\| \sum_{\eta\in\Xi_{M'}} \sum_{{}^\eta\Gamma/{}^\eta\Gamma\cap P_o} e^{(\lambda+\Lambda_R'-\varrho_o)(H_o(x\eta^{-1}\gamma))} \qquad \text{(see Lemma 103)},$$

and Lemma 109 follows from the results of Chap.II,§2.

Using the same notations we define $D(P|P')$ as the set of all points Λ_o of $\mathfrak{a}_C^*$ which have a compact neighbourhood Ω in $\mathfrak{a}_C^*$ with the following properties:

1) ${}^*\Omega \subset D({}^*P|{}^*A)$.

2) For any compact subset C of G and any $\varphi \in L_P(\sigma,\xi_P)$ the series

$$\sum_{\Gamma/\Gamma\cap P'} |E({}^*P|{}^*A:{}^*\Lambda:\varphi:x\gamma)\, e^{(\Lambda-\varrho)(H'(x\gamma))}|$$

converges uniformly for $\Lambda \in \Omega$ and $x \in C$; there exist c and N such that

$$\sum_{\Gamma/\Gamma\cap P'} |E(^*P|^*A:{}^*\Lambda:\varphi:x\gamma)\, e^{(\Lambda-\varrho)(H'(x\gamma))}| \leq c\|\varphi\|\ \|x\|^N$$

for all $\Lambda \in \Omega$, $\varphi \in L_P(\sigma,\xi_P)$, $x \in G$.

Lemma 109 implies that, given a compact set $^*\omega$ in $D(^*P|^*A)$, we can choose $\Lambda_o' \in \mathfrak{a}'^*$ such that $\Lambda_o' - Cl(\mathfrak{a}_c'^*)^+ + {}^*\omega \subset D(P|P')$. In particular $^*D(P|P') = D(^*P|^*A)$. It is clear that $D(P|P')$ is open in $\mathfrak{m}_c^*$ and contains $(\mathfrak{a}_c^*)^-$ (cf. Lemma 105). Moreover $D(P|P')$ is connected. To see this, suppose first that $\Lambda_1,\Lambda_2 \in D(P|P')$ and $^*\Lambda_1 = {}^*\Lambda_2$. From the trivial inequality

$$|e^{(t\Lambda_1+(1-t)\Lambda_2-\varrho)(H'(x\gamma))}| \leq t|e^{(\Lambda_1-\varrho)(H'(x\gamma))}| + (1-t)|e^{(\Lambda_2-\varrho)(H'(x\gamma))}| \quad (0 \leq t \leq 1)$$

one sees that the line segment joining Λ_1 and Λ_2 lies in $D(P|P')$. Now let Λ_1,Λ_2 be arbitrary points in $D(P|P')$. By the induction hypothesis $D(^*P|^*A)$ is connected, so we can choose a compact connected set $^*\omega$ in $D(^*P|^*A)$ such that $^*\Lambda_i \in {}^*\omega$ $(i = 1,2)$. Choose $\Lambda_o' \in \mathfrak{a}'^*$ such that $\Lambda_o' + {}^*\omega \subset D(P|P')$. We know already that Λ_i can be connected to $\Lambda_o' + {}^*\Lambda_i$ in $D(P|P')$. Since $^*\Lambda_1$ and $^*\Lambda_2$ can be connected in $^*\omega$, $\Lambda_o' + {}^*\Lambda_1$ and $\Lambda_o' + {}^*\Lambda_2$ can be connected in $\Lambda_o' + {}^*\omega$. So Λ_1 and Λ_2 can be connected in $D(P|P')$.

$\sum_{\Gamma/\Gamma\cap P'} E(^*P|^*A:{}^*\Lambda:\varphi:x\gamma)\, e^{(\Lambda-\varrho)(H'(x\gamma))}$ is continuous on $D(P|P') \times G$ and is an extension of $E(P|A:\Lambda:\varphi:x)$ (Lemma 105).

In the rest of this paragraph we use the notation of Theorem 8.

Fix i $(1 \leq i \leq r)$ and let (P',A') be a cuspidal pair of G which dominates (P_i,A_i) with rank $P' \geq 1$. Put $(^*P_i,{}^*A_i) = (M' \cap P_i,\ M' \cap A_i)$, $(P_{ik}',A_{ik}') = {}^{y_{ik}}(P',A')$, $(^*P_{ik},{}^*A_{ik}) = (M_{ik}' \cap P_{ik},\ M_{ik}' \cap A_{ik})$ $(1 \leq k \leq r_i)$; $(^*P_i,{}^*A_i)$ is a cuspidal pair of M' with rank $^*P_i < q$, (P_{ik}',A_{ik}') is a cuspidal pair of G which dominates (P_{ik},A_{ik}) and $(^*P_{ik},{}^*A_{ik})$ is a cuspidal pair of M_{ik}' with rank $^*P_{ik} = \text{rank}\ {}^*P_i < q$. Define

$$\underline{D}(P_i|P') = \bigcap_{1\leq k\leq r_i} {}^{y_{ik}^{-1}} D(P_{ik}|P_{ik}') \ .$$

Then $\mathfrak{D}(P_i|P')$ is a subset of $\mathfrak{a}^*_{i,C}$, which is obviously open and contains $(\mathfrak{a}^*_{i,C})^-$. Moreover it is connected, as we shall prove. Put

$$\mathfrak{D}(^*P_i|^*A_i) = \bigcap_{1\leq k\leq r_i} {}^{y_{ik}^{-1}}D(^*P_{ik}|^*A_{ik}) .$$

This is a subset of $^*\mathfrak{a}^*_{i,C}$. From Lemma 109 we see immediately that, given a compact subset $^*\omega$ of $\mathfrak{D}(^*P_i|^*A_i)$, we can choose $\Lambda_o' \in \mathfrak{a}'^*$ such that $\Lambda_o' - Cl(\mathfrak{a}_c'^*)^+ + {}^*\omega \subset \mathfrak{D}(P_i|P')$. In particular $^*\mathfrak{D}(P_i|P') = \mathfrak{D}(^*P_i|^*A_i)$. If $\Lambda_1,\Lambda_2 \in \mathfrak{D}(P_i|P')$ and $^*\Lambda_1 = {}^*\Lambda_2$, then the line segment joining Λ_1 and Λ_2 lies in $\mathfrak{D}(P_i|P')$, because the line segment joining $^{y_{ik}}\Lambda_1$ and $^{y_{ik}}\Lambda_2$ lies in $D(P_{ik}|P_{ik}')$ for $1 \leq k \leq r_i$. From the induction hypothesis follows that $\mathfrak{D}(^*P_i|^*A_i)$ is connected. One proves now in the same way as for $D(P|P')$ (see above) that $\mathfrak{D}(P_i|P')$ is connected.

The function

$$\mathfrak{E}(P_i|A_i:\Lambda:\varphi:x) = \sum_{1\leq k\leq r_i} E(P_{ik}|A_{ik}:{}^{y_{ik}}\Lambda:\varphi_k:x) \qquad (\varphi = (\varphi_k) \in \mathfrak{L}_i)$$

is continuous on $\mathfrak{D}(P_i|P') \times G$ and holomorphic as a function of Λ for $\Lambda \in \mathfrak{D}(P_i|P')$.

Fix i,j $(1 \leq i, j \leq r)$. Suppose (P',A') is a cuspidal pair such that $(P',A') \succ (P_i,A_i)$, $(P',A') \succ (P_j,A_j)$, rank $P' \geq 1$. Let $w'(\mathfrak{a}_i,\mathfrak{a}_j)$ be the set of all $s \in w(\mathfrak{a}_i,\mathfrak{a}_j)$ such that $^sH = H$ for $H \in \mathfrak{a}'$.

Lemma 110. If $s \in w'(\mathfrak{a}_i,\mathfrak{a}_j)$, then $c_{ji}(s:\Lambda)$ is a meromorphic function of Λ on $\mathfrak{a}^*_{i,C}$ whose poles lie along hyperplanes.

Proof. We have to prove that $c_{P_{jl}|P_{ik}}(y_{jl}sy_{ik}^{-1}:{}^{y_{ik}}\Lambda)$ is a meromorphic function of Λ $(1 \leq k \leq r_i, 1 \leq l \leq r_j)$. By the corollary of Lemma 106 this function is identically zero unless $P_{ik}' = {}^{y_{ik}}P'$ and $P_{jl}' = {}^{y_{jl}}P'$ are conjugated under Γ . Suppose $^{\gamma}P'_{jl} = P'_{ik}$ with $\gamma \in \Gamma$ and choose $u \in (U'_{jl})_{\mathbf{Q}}$ such that $^{\gamma u}A'_{jl} = A'_{ik}$ (here we have put $A'_{ik} = {}^{y_{ik}}A'$, $A'_{jl} = {}^{y_{jl}}A'$). Replacing y_{jl}, P_{jl}, A_{jl} by $\gamma u y_{jl}$, $^{\gamma}P_{jl}$, $^{\gamma u}A_{jl}$ (which is allowed by Lemma 102) we get $P'_{jl} = P'_{ik}$, $A'_{jl} = A'_{ik}$ and $y_{jl}sy_{ik}^{-1}$ leaves $\mathfrak{a}'_{ik}$ pointwise fixed. Call $y_{jl}sy_{ik}^{-1} = s_o$ and $^{y_{ik}}\Lambda = \Lambda_o$; then, by Lemma 108,

$c_{P_{jl}|P_{ik}}(s_o:\Lambda_o) = c_{*P_{jl}|*P_{ik}}(*s_o:*\Lambda_o)$ and it follows from the induction hypothesis that this is a meromorphic function of Λ with its poles along hyperplanes.

Lemma 111. Fix $s \in w'(\mathfrak{a}_i,\mathfrak{a}_j)$ and let $\mathfrak{D}_{ij}(s)$ be the set of all Λ_o in $\mathfrak{D}(P_i|P') \cap s^{-1}\mathfrak{D}(P_j|P')$ such that $c_{ji}(s:\Lambda)$ is holomorphic at Λ_o. Then $\mathfrak{D}_{ij}(s)$ is not empty, and

$$\mathfrak{E}(P_j|A_j:s\Lambda:c_{ji}(s:\Lambda)\varphi) = \mathfrak{E}(P_i|A_i:\Lambda:\varphi)$$

for $\varphi \in \mathfrak{L}_i$, $\Lambda \in \mathfrak{D}_{ij}(s)$.

Proof. Put, as before, $*P_i = M' \cap P_i$, $P'_{ik} = {}^{y_{ik}}P'$, etc.

We know from the induction hypothesis that $\mathfrak{D}(*P_i|*A_i) \cap s^{-1}\mathfrak{D}(*P_j|*A_j)$ is open and not empty and from Lemma 109 that, given a compact subset $*\omega$ of that set, we can choose $\Lambda'_o \in \mathfrak{a}'^*$ such that $\Lambda'_o - Cl(\mathfrak{a}'^*_c)^+ + *\omega \subset \mathfrak{D}(P_i|P') \cap s^{-1}\mathfrak{D}(P_j|P')$. Together with Lemma 110 this implies that $\mathfrak{D}_{ij}(s)$ is not empty.

To prove the second assertion of Lemma 111 we fix k, $1 \le k \le r_i$, and take $\varphi \in L_{ik}$. We assume that $i = 1$, $j = 2$, $k = 1$ (the case that $i = j$ is not lost in this way). We have to prove that

$$\sum_{1\le l\le r_2} E(P_{2l}|A_{2l}:{}^{y_{2l}}(s\Lambda):c_{P_{2l}|P_{11}}(y_{2l}sy_{11}^{-1}:{}^{y_{11}}\Lambda)\varphi) = E(P_{11}|A_{11}:{}^{y_{11}}\Lambda:\varphi) .$$

The problem is unchanged when we replace

$$(P_1,A_1),\quad y_{1k},\quad (P',A'),\quad (P_2,A_2),\quad y_{2l},\quad s,\quad \Lambda$$

by $$(P_{11},A_{11}),\quad y_{1k}y_{11}^{-1},\quad (P'_{11},A'_{11}),\quad {}^{y_{11}}(P_2,A_2),\quad y_{2l}y_{11}^{-1},\quad y_{11}sy_{11}^{-1},\quad {}^{y_{11}}\Lambda .$$

Then we get $y_{11} = 1$ and we have to prove

$$\sum_{1\le l\le r_2} E(P_{2l}|A_{2l}:{}^{y_{2l}}(s\Lambda):c_{P_{2l}|P_1}(y_{2l}s:\Lambda)\varphi) = E(P_1|A_1:\Lambda:\varphi) .$$

Using Lemma 102 one sees that the problem is unchanged when we replace y_{2l} by $\gamma_l u_l y_{2l}$ with $\gamma_l \in \Gamma$, $u_l \in (U'_{2l})_{\mathbb{Q}}$ $(1 \le l \le r_2)$. When the γ_l and u_l are appropriately chosen we get: $(P'_{2l},A'_{2l}) = (P',A')$ whenever P'_{2l} is conjugate to P' under Γ. Assume that $(P'_{2l},A'_{2l}) = (P',A')$ for $1 \le l \le *r_2$ and that P'_{2l} is not conjugate to

P' under Γ for $l > {}^*r_2$. Now $y_{2l} \in M'A'$ for $1 \leq l \leq {}^*r_2$. The groups P_{2l} $(1 \leq l \leq {}^*r_2)$ form a complete set of representatives for the groups which are conjugated to P_2 under $P'_{\mathbb{Q}}$ modulo conjugation under $\Gamma \cap P'$ (if $y \in P'_{\mathbb{Q}}$, we can choose γ and l with $\gamma \in \Gamma$, $1 \leq l \leq r_2$ such that ${}^yP_2 = {}^{\gamma}P_{2l}$; then $\gamma y_{2l} \in yP_2 \subset P'$, so ${}^{\gamma}P'_{2l} = P'$; hence $l \leq {}^*r_2$ and $\gamma \in P'$). So the groups ${}^*P_{2l} = M' \cap P_{2l}$ $(1 \leq l \leq {}^*r_2)$ form a complete set of representatives for the cuspidal subgroups of M' which are conjugate (under $M'_{\mathbb{Q}}$) to ${}^*P_2 = M' \cap P_2$ modulo conjugation under $\Gamma_{M'}$, as follows immediately from the following lemma, which is easy to prove.

<u>Lemma 112</u>. Let (P_1,A_1), (P_2,A_2) and (P',A') be cuspidal pairs of G such that $(P',A') > (P_i,A_i)$ for $i = 1,2$. Define cuspidal pairs $({}^*P_i,{}^*A_i) = (M' \cap P_i, M' \cap A_i)$ in M'. Then:

1) ${}^*P_1$ and ${}^*P_2$ are conjugate under $M'_{\mathbb{Q}}$ if and only if P_1 and P_2 are conjugate under $P'_{\mathbb{Q}}$.

2) ${}^*P_1$ and ${}^*P_2$ are conjugate under $\Gamma_{M'}$ if and only if P_1 and P_2 are conjugate under $\Gamma \cap P'$.

3) $({}^*P_1,{}^*A_1)$ and $({}^*P_2,{}^*A_2)$ are conjugate under $M'_{\mathbb{Q}}$ if and only if (P_1,A_1) and (P_2,A_2) are conjugate under $P'_{\mathbb{Q}}$.

The cuspidal pairs $({}^*P_2,{}^*A_2) = (M' \cap P_2, M' \cap A_2)$ and $({}^*P_{2l},{}^*A_{2l}) = (M' \cap P_{2l}, M' \cap A_{2l})$ (l fixed and $1 \leq l \leq {}^*r_2$) are conjugate under $M'_{\mathbb{Q}}$ (Lemma 112,3)). Choose $m_{2l} \in M'_{\mathbb{Q}}$ such that $({}^*P_{2l},{}^*A_{2l}) = {}^{m_{2l}}({}^*P_2,{}^*A_2)$ $(1 \leq l \leq {}^*r_2)$. Then $(P_{2l},A_{2l}) = {}^{m_{2l}}(P_2,A_2)$ (Lemma 2), so $m_{2l}^{-1}\, y_{2l} \in M_2A_2$ and $\mathrm{Ad}(y_{2l})s = \mathrm{Ad}(m_{2l})s$ on $\mathfrak{a}_1$. In view of Lemma 108 we have $c_{P_{2l}|P_1}(y_{2l}s:\Lambda) = c_{P_{2l}|P_1}(m_{2l}s:\Lambda) = c_{{}^*P_{2l}|{}^*P_1}({}^*(m_{2l}s):{}^*\Lambda) = c_{{}^*P_{2l}|{}^*P_1}(m_{2l}{}^*s:{}^*\Lambda)$ for $1 \leq l \leq {}^*r_2$; and, by the corollary of Lemma 106, $c_{P_{2l}|P_1}(y_{2l}s:\Lambda) = 0$ for $l > {}^*r_2$. After these remarks we see, using definitions only:

$$\sum_{1 \leq l \leq r_2} E(P_{2l} | A_{2l} : {}^{y_{2l}}(s\Lambda) : c_{P_{2l}|P_1}(y_{2l}s:\Lambda)\varphi : x) =$$

$$\sum_{1 \leq l \leq {}^*r_2} E(P_{2l} | A_{2l} : {}^{y_{2l}}(s\Lambda) : c_{{}^*P_{2l}|{}^*P_1}(m_{2l}{}^*s:{}^*\Lambda)\varphi : x) =$$

$$\sum_{1 \leq l \leq {}^*r_2} \sum_{\Gamma/\Gamma\cap P'} E({}^*P_{2l} | {}^*A_{2l} : {}^{m_{2l}}({}^*s{}^*\Lambda) : c_{{}^*P_{2l}|{}^*P_1}(m_{2l}{}^*s:{}^*\Lambda)\varphi : x\gamma) e^{(s\Lambda-\varrho_2)(H'(x\gamma))} =$$

$$\sum_{\Gamma/\Gamma\cap P'} \underset{\sim}{E}({}^*P_2 | {}^*A_2 : {}^*s{}^*\Lambda : c_{21}({}^*s:{}^*\Lambda)\varphi : x\gamma) \ e^{(s\Lambda-\varrho_2)(H'(x\gamma))} .$$

By the induction hypothesis we have

$$\underset{\sim}{E}({}^*P_2 | {}^*A_2 : {}^*s{}^*\Lambda : c_{21}({}^*s:{}^*\Lambda)\varphi) = \underset{\sim}{E}({}^*P_1 | {}^*A_1 : {}^*\Lambda : \varphi) .$$

On the other hand $s\Lambda - \varrho_2 = \Lambda - \varrho_1$ on $\mathfrak{a}'$. So we get

$$\sum_{\Gamma/\Gamma\cap P'} E({}^*P_1 | {}^*A_1 : {}^*\Lambda : \varphi : x\gamma) \ e^{(\Lambda-\varrho_1)(H'(x\gamma))} = E(P_1 | A_1 : \Lambda : \varphi : x) .$$

The proof of Lemma 111 is finished.

Put, under the assumptions of Lemma 111,

$$f = \underset{\sim}{E}(P_j | A_j : s\Lambda : c_{ji}(s:\Lambda)\varphi) = \underset{\sim}{E}(P_i | A_i : \Lambda : \varphi) .$$

Then, for $1 \leq k \leq r$, $1 \leq l \leq r_k$:

$$f_{P_{kl}}(x) = \sum_{t \in w(\mathfrak{a}_i, \mathfrak{a}_k)} (c_{ki}(t:\Lambda)\varphi)_{kl}(x) e^{{}^{y_{kl}}(t\Lambda-\varrho_k)(H_{kl}(x))} ,$$

and also

$$f_{P_{kl}}(x) = \sum_{t \in w(\mathfrak{a}_j, \mathfrak{a}_k)} (c_{kj}(t:s\Lambda)c_{ji}(s:\Lambda)\varphi)_{kl}(x) \ e^{{}^{y_{kl}}(ts\Lambda-\varrho_k)(H_{kl}(x))} .$$

This gives:

<u>Lemma 113</u>. Fix $s \in w'(\mathfrak{a}_i, \mathfrak{a}_j)$. Then $c_{ki}(ts:\Lambda) = c_{kj}(t:s\Lambda)c_{ji}(s:\Lambda)$ for any k $(1 \leq k \leq r)$, $t \in w(\mathfrak{a}_j, \mathfrak{a}_k)$, $\Lambda \in \underset{\sim}{D}(P_i | P') \cap s^{-1} \underset{\sim}{D}(P_j | P')$.

Lemma 114. Fix i,j $(1 \leq i, j \leq r)$. Let (P',A') be a cuspidal pair of G and y an element of $G_{\mathbb{Q}}$ such that rank $P' \geq 1$ and $(P',A') \succ (P_i,A_i)$ and ${}^{y}(P',A') \succ (P_j,A_j)$. Let s be an element of $w(\mathfrak{a}_i,\mathfrak{a}_j)$ such that $s = \mathrm{Ad}(y)$ on $\mathfrak{a}'$. Then the set $\underline{\mathfrak{D}}(P_i|P') \cap s^{-1}\,\underline{\mathfrak{D}}(P_j|{}^{y}P')$ is not empty and

$$c_{ki}(ts:\Lambda) = c_{kj}(t:s\Lambda)c_{ji}(s:\Lambda)$$

for $1 \leq k \leq r$, $t \in w(\mathfrak{a}_j,\mathfrak{a}_k)$, $\Lambda \in \underline{\mathfrak{D}}(P_i|P') \cap s^{-1}\,\underline{\mathfrak{D}}(P_j|{}^{y}P')$.

Proof. Put $(P_j',A_j') = {}^{y^{-1}}(P_j,A_j)$, $y_{jl}' = y_{jl}y$ $(1 \leq l \leq r_j)$, $s' = y^{-1}s$ and $t' = ty$. Then $s^{-1}\,\underline{\mathfrak{D}}(P_j|{}^{y}P') = s'^{-1}\,\underline{\mathfrak{D}}(P_j'|P')$. Apply Lemmas 111 and 113 to P_i, P_j' etc.

§4. End of the proof of Theorem 8.

First we mention some well known facts concerning Weyl chambers.

Let A be a split component of some cuspidal group in $\mathcal{C}$. Call $\mathcal{P}_A$ the set of all cuspidal subgroups of G which have A as a split component. Let Σ be the set of all roots of $(\mathfrak{g},\mathfrak{a})$ and for $\alpha \in \Sigma$, let Σ_α^+ be the set of all elements of Σ of the form $k\alpha$ with $k \in \mathbb{R}$, $k > 0$ and put $\Sigma_\alpha = \Sigma_\alpha^+ \cup \Sigma_{-\alpha}^+$. For $H \in \mathfrak{a}$, let $\Sigma(H)$ denote the set of all $\alpha \in \Sigma$ such that $\alpha(H) = 0$. By definition H is <u>regular</u> in $\mathfrak{a}$ if and only if $\Sigma(H)$ is empty and H is <u>semi-regular</u> in $\mathfrak{a}$ if and only if $\Sigma(H) = \Sigma_\alpha$ for some $\alpha \in \Sigma$. The subset of $\mathfrak{a}$ consisting of the regular and the semi-regular elements is connected. The connected components of the set of all regular elements of $\mathfrak{a}$ are called (Weyl) chambers. We denote the set of all chambers by $\mathcal{L}$. We define:

if $C \in \mathcal{L}$, then $\Sigma(C) = \{\alpha \in \Sigma : \alpha(H) > 0 \text{ for } H \in C\}$;

if $H \in \mathfrak{a}$, then $\mathcal{L}(H) = \{C \in \mathcal{L} : H \in \mathrm{Cl}(C)\}$.

Then, for $C \in \mathcal{L}$, we have $C = \{H \in \mathfrak{a} : \alpha(H) > 0 \text{ for } \alpha \in \Sigma(C)$; and, if $H \in \mathfrak{a}$, $C \in \mathcal{L}$, then $C \in \mathcal{L}(H)$ if and only if $\Sigma(C)$ contains all $\alpha \in \Sigma$ such that $\alpha(H) > 0$. For $C \in \mathcal{L}$, put $\mathfrak{n}_C = \sum_{\alpha \in \Sigma(C)} \mathfrak{g}_\alpha$ ($\mathfrak{g}_\alpha$ = root space corresponding to α), $U_C = \exp \mathfrak{n}_C$, $P_C = MAU_C$ (MA = centralizer of A in G, M is as always). Then $C \longmapsto P_C$ is a bijective mapping of $\mathcal{L}$ on $\mathcal{P}_A$, which is compatible with the action of $w(\mathfrak{a})$. If $s \in w(\mathfrak{a})$, $s \neq 1$, then $sC \neq C$ for any $C \in \mathcal{L}$, hence ${}^sP \neq P$ for any $P \in \mathcal{P}_A$. On the other hand, two elements of $\mathcal{P}_A$ are conjugate under $G_{\mathbb{Q}}$ if and only if they are conjugate under $w(\mathfrak{a})$, and each element of $\mathcal{C}$ is conjugate under $G_{\mathbb{Q}}$ to some element of $\mathcal{P}_A$, so we have bijections

$$\mathcal{C}/G_{\mathbb{Q}} \leftrightarrow \mathcal{P}_A/w(\mathfrak{a}) \leftrightarrow \mathcal{L}/w(\mathfrak{a}) .$$

Let (P_o, A_o) be a mincuspidal pair and let $\mathfrak{a}_o^+$ be the set of all $H \in \mathfrak{a}_o$ such that $\alpha(H) > 0$ for every $\alpha \in \Sigma(P_o|A_o)$. Then every point of $\mathfrak{a}_o$ is conjugate under $G_{\mathbb{Q}}$ to exactly one point of $Cl(\mathfrak{a}_o^+)$; two points of $\mathfrak{a}_o$ are conjugate under $G_{\mathbb{Q}}$ if and only if they are conjugate under $w(\mathfrak{a}_o)$; $w(\mathfrak{a}_o)$ acts simply transitively on the set $\mathcal{L}_o$ of chambers of $\mathfrak{a}_o$.

Returning to the preceding notations we have, for $C \in \mathcal{L}$, $\Sigma(C) = \Sigma(P_C|A)$ and we define $\Sigma^o(C) = \Sigma^o(P_C|A)$.

Two chambers C_1, C_2 are called <u>adjacent</u> if 1° $C_1 \neq C_2$ and 2° $Cl(C_1) \cap Cl(C_2)$ contains a semi-regular element. Fix $C \in \mathcal{L}$ and $\alpha \in \Sigma^o(C)$. Then there is a unique chamber C^α adjacent to C such that $-\alpha \in \Sigma(C^\alpha)$. In fact $\Sigma(C^\alpha) = -\Sigma_\alpha^+ \cup \Sigma_\alpha'(C)$ where $\Sigma_\alpha'(C)$ is the complement of Σ_α^+ in $\Sigma(C)$. Conversely, if C' is any chamber adjacent to C, then there is a unique $\alpha \in \Sigma^o(C)$ such that $C' = C^\alpha$. From this we see that, if H is semi-regular, then $\mathcal{L}(H)$ consists of two elements.

Given $C, C' \in \mathcal{L}$ we can find a sequence $C_1, \ldots, C_N$ of chambers such that

1) $C_1 = C$, $C_N = C'$;
2) C_i, C_{i+1} are adjacent $(1 \leq i < N)$;
3) if α_i is the element of $\Sigma^o(C_i)$ such that $C_{i+1} = C_i^{\alpha_i}$ $(1 \leq i < N)$, then $\alpha_i \in \Sigma(C_k)$ for $k \leq i$ and $-\alpha_i \in \Sigma(C_k)$ for $k > i$.

Fix $C \in \mathcal{L}$ and put $P = P_C$, $U = U_C$, so $P = MAU$. Fix $\alpha \in \Sigma^o(C)$ and put $P^\alpha = P_{C^\alpha}$, $U^\alpha = U_{C^\alpha}$, so $P^\alpha = MAU^\alpha$. Put $(P',A') = (P,A)_{\{\alpha\}}$ (notation of Chap.I,§1). Then rank $P' = q-1 \geq 1$ and $(P',A') \succ (P,A)$ and $(P',A') \succ (P^\alpha,A)$.

Let $C_1, \ldots, C_r$ be the distinct chambers of $\mathfrak{a}$. Put $P_i = P_{C_i}$ $(1 \leq i \leq r)$. We shall prove Theorem 8 for these groups P_i. In the general case the theorem results trivially from this particular case.

Lemma 115. Assume $1 \leq i, j \leq r$ and $s \in w(\mathfrak{a})$ such that $sC_i = C_j$. Then $c_{ji}(s:\Lambda)$ is an entire function of Λ on $\mathfrak{a}_C^*$.

Proof. If $i = j$, then $s = 1$ and $c_{ii}(1:\Lambda) = 1$. Assume $i = 1, j = 2$. Fix k and l with $1 \leq k \leq r_1$, $1 \leq l \leq r_2$ and put $s' = y_{2l} s y_{1k}^{-1}$, $\Lambda' = {}^{y_{1k}}\Lambda$. We have to prove that $c_{P_{2l}|P_{1k}}(s':\Lambda')$ is an entire function of Λ. Fix $y \in G_{\mathbb{Q}}$ such that $\mathrm{Ad}(y) = s$ on $\mathfrak{a}$; put $y' = y_{2l} y y_{1k}^{-1}$, then $\mathrm{Ad}(y') = s'$ on $\mathfrak{a}_{1k}$. Since $(P_{2l}, A_{2l}) = {}^{y'}(P_{1k}, A_{1k})$, we have $c_{P_{2l}|P_{1k}}(s':\Lambda') = 0$ unless P_{1k} and P_{2l} are conjugate under Γ (Corollary of Lemma 106). Suppose $P_{2l} = {}^{\gamma}P_{1k}$, $\gamma \in \Gamma$; choose $u \in (U_{1k})_{\mathbb{Q}}$ such that $A_{2l} = {}^{\gamma u}A_{1k}$. Then $\mathrm{Ad}(\gamma u) = \mathrm{Ad}(y') = s'$ on $\mathfrak{a}_{1k}$. So $c_{P_{2l}|P_{1k}}(s':\Lambda') = c_{P_{2l}|P_{1k}}(\gamma u:\Lambda') = e^{(\Lambda'-\varrho_{1k})(H_{1k}(\gamma))} \tau_\gamma\, c_{P_{1k}|P_{1k}}(1:\Lambda') =$
$= e^{{}^{y_{1k}}(\Lambda-\varrho_1)(H_{1k}(\gamma))} \tau_\gamma$ (Lemma 102).

Lemma 116. Let C_i, C_j be two adjacent chambers.

1) $c_{ji}(1:\Lambda)$ is a meromorphic function of Λ on $\mathfrak{a}_C^*$ whose poles lie along hyperplanes.

2) Let α be the root in $\Sigma^o(C_i)$ such that $C_j = C_i^\alpha$. Let D_α be the set of all $\Lambda \in \mathfrak{a}_C^*$ such that $\mathrm{Re}\langle\Lambda,\alpha\rangle \leq 0$. There exists a finite set $\{b_1,\dots,b_N\}$ of real numbers such that $b_1 < b_2 < \dots < b_N < 0$ with the following property. Put $p(\Lambda) =$
$= \prod_{1 \leq i \leq N} (\langle\Lambda,\alpha\rangle - b_i)$. Then $p(\Lambda)\, c_{ji}(1:\Lambda)$ is holomorphic on D_α.

Proof. Assume $i = 1$, $j = 2$, so $C_2 = C_1^\alpha$, $\alpha \in \Sigma^o(C_1)$. Define $(P',A') =$
$= (P_1,A)_{\{\alpha\}}$; then (P',A') dominates both (P_1,A) and (P_2,A). Put $(P'_{1k},A'_{1k}) =$
$= {}^{y_{1k}}(P',A')$, $(P'_{2l},A'_{2l}) = {}^{y_{2l}}(P',A')$ $(1 \leq k \leq r_1,\ 1 \leq l \leq r_2)$. By the corollary of Lemma 106, $c_{P_{2l}|P_{1k}}(y_{2l}y_{1k}^{-1}: {}^{y_{1k}}\Lambda) = 0$ unless P'_{1k} and P'_{2l} are conjugate under Γ. Fix k, l and assume that $P'_{1k} = {}^{\gamma}P'_{2l}$, $\gamma \in \Gamma$. Choose $u \in (U'_{2l})_{\mathbb{Q}}$ such that ${}^{\gamma u}A'_{2l} = A'_{1k}$. Define $s \in w(\mathfrak{a}_{1k}, \mathfrak{a}_{2l})$ by $s = \mathrm{Ad}(y_{2l}y_{1k}^{-1})$ on $\mathfrak{a}_{1k}$ and put $\Lambda_o = {}^{y_{1k}}\Lambda$. By Lemma 102 we have

$$(*)\qquad c_{P_{21}|P_{1k}}(s:\Lambda_o) = e^{-{}^{y_{21}}(\Lambda-\varrho_2)(H_{21}(\gamma))}\,\tau_\gamma^{-1}\, c_{{}^{\gamma}P_{21}|P_{1k}}(\gamma us:\Lambda_o)\ .$$

Put $(P'',A'') = (P_{1k}',A_{1k}')$; (P'',A'') dominates both (P_{1k},A_{1k}) and ${}^{\gamma u}(P_{21},A_{21})$. Put $({}^*P_1,{}^*A_1) = (M'' \cap P_{1k}\ ,\ M'' \cap A_{1k})$, $({}^*P_2,{}^*A_2) = (M'' \cap {}^{\gamma u}P_{21},\ M'' \cap {}^{\gamma u}A_{21})$. Since $\gamma us = \mathrm{Ad}(\gamma u\ y_{21}y_{1k}^{-1}) = 1$ on $\mathfrak{a}''$, we have by Lemma 108:

$$c_{{}^{\gamma}P_{21}|P_{1k}}(\gamma us:\Lambda_o) = c_{{}^*P_2|{}^*P_1}({}^*(\gamma us):{}^*\Lambda_o) = c_{{}^*P_2|{}^*P_1}({}^*s_o:{}^*\Lambda_o)\ ,$$

where $s_o = \gamma us$. Now ${}^*P_1$ and ${}^*P_2$ have rank 1 . From Chapter IV we know that $c_{{}^*P_2|{}^*P_1}({}^*s_o:{}^*\Lambda_o)$ is a meromorphic function of ${}^*\Lambda_o$ on ${}^*\mathfrak{a}_{1,C}^*$; it is holomorphic when $\mathrm{Re}\langle{}^*\Lambda_o,{}^*\alpha_o\rangle \leq 0$ except for a finite number of simple poles which occur at points where $\langle{}^*\Lambda_o,{}^*\alpha_o\rangle$ is real and strictly negative (here we put $\alpha_o = {}^{y_{1k}}\alpha$). Now ${}^*\Lambda_o = \langle\Lambda_o,\alpha_o\rangle\langle\alpha_o,\alpha_o\rangle^{-1}\,{}^*\alpha_o$ and $\langle\Lambda_o,\alpha_o\rangle = \langle\Lambda,\alpha\rangle$, so $\langle{}^*\Lambda_o,{}^*\alpha_o\rangle = c\langle\Lambda,\alpha\rangle$ where c is a constant, $c > 0$. So the last factor in the right hand side of $(*)$ is a function of $\langle\Lambda,\alpha\rangle$ and all assertions of Lemma 116 are proved now.

<u>Corollary</u>. Let p be as in Lemma 116,2) (so that $p(\Lambda)c_{ji}(1:\Lambda)$ is holomorphic on D_α). If ω is a compact subset of $D_\alpha \cap \mathfrak{a}^*$, we can choose c such that

$$|p(\Lambda)c_{ji}(1:\Lambda)| \leq c(1 + |\Lambda_I|)^N \quad \text{for } \Lambda = \Lambda_R + i\Lambda_I \text{ with } \Lambda_R \in \omega\ .$$

<u>Proof</u>. In view of the proof of Lemma 116 this follows from Lemma 101.

Now we consider $c_{ji}(s:\Lambda)$ for arbitrary i, j and s . Let m be the index such that $sC_i = C_m$. Then, by Lemma 114 ,

$$(**)\qquad c_{ji}(s:\Lambda) = c_{jm}(1:s\Lambda)\ c_{mi}(s:\Lambda)$$

on some non empty open subset of $\mathfrak{a}_C^*$ (take in Lemma 114: $i = i$, $j = m$, $(P',A') = (P_i,A)$, $y \in G_{\mathbb{Q}}$ such that $\mathrm{Ad}(y) = s$ on $\mathfrak{a}$). Since $c_{mi}(s:\Lambda)$ is an entire function of Λ (Lemma 115), it is enough to consider $c_{ji}(1:\Lambda)$ for arbitrary i and j .

Fix i, j and choose a sequence $C_{i_1},\ldots,C_{i_N}$ of chambers satisfying the conditions 1), 2), 3) in the first part of this paragraph with $C = C_i$, $C' = C_j$. To simplify the notations we shall assume that $i_k = k$ $(1 \leq k \leq N)$. Thus we consider $c_{N1}(1:\Lambda)$ and the sequence $C_1, \ldots, C_N$ is such that C_i, C_{i+1} are adjacent $(1 \leq i < N)$ and, if $C_{i+1} = C_i^{\alpha_i}$ with $\alpha_i \in \Sigma^o(C_i)$, then $\alpha_i \in \Sigma(C_k)$ for $k \leq i$ and $-\alpha_i \in \Sigma(C_k)$ for $k > i$. Put $c_i(\Lambda) = c_{i+1,i}(1:\Lambda)$ $(1 \leq i < N)$; then c_i is meromorphic on $\mathfrak{a}_C^*$ (Lemma 116).

<u>Lemma 117</u>. $c_{N1}(1:\Lambda)$ is meromorphic on $\mathfrak{a}_C^*$ and its poles lie along hyperplanes. In fact,

$$c_{N1}(1:\Lambda) = c_{N-1}(\Lambda) \ldots c_2(\Lambda)\, c_1(\Lambda) .$$

<u>Proof</u>. It is enough to prove that

$$c_{N,N-k}(1:\Lambda) = c_{N-1}(\Lambda) \ldots c_{N-k}(\Lambda) \qquad \text{for } 0 \leq k \leq N-1 .$$

This equality holds for $k = 0$. Suppose it holds for some k, $0 \leq k \leq N-2$. By Lemma 113 we have $c_{N,N-k-1}(1:\Lambda) = c_{N,N-k}(1:\Lambda)c_{N-k-1}(\Lambda)$ on some non empty open subset of $\mathfrak{a}_C^*$. Then the equality holds for $k + 1$.

Moreover, we can apply Lemma 116, 2) and the corollary of Lemma 116 to the functions $c_i(\Lambda)$ $(1 \leq i < N)$. Then we return to the general situation and to the formula (**). Observe that $c_{mi}(s:\Lambda)$ is given by exponential functions (see the proof of Lemma 115). Thus we get:

<u>Lemma 118</u>. Fix i,j $(1 \leq i, j \leq r)$ and $s \in w(\mathfrak{a})$.

1) $c_{ji}(s:\Lambda)$ is a meromorphic function of Λ on $\mathfrak{a}_C^*$ whose poles lie along hyperplanes.

2) There exist roots $\alpha_1, \ldots, \alpha_N$ in $\Sigma(P_i|A)$ no two of which are proportional and real numbers b_{kl} $(1 \leq k \leq N,\ 1 \leq l \leq p_k)$ such that $b_{k1} < b_{k2} < \ldots < b_{kp_k}$ with the following properties. Put

$$p(\Lambda) = \prod_{1\leq k\leq N} \ \prod_{1\leq l\leq p_k} (\langle\Lambda,\alpha_k\rangle - b_{kl}) .$$

Then $p(\Lambda)c_{ji}(s:\Lambda)$ is holomorphic on $-\mathrm{Cl}(\mathfrak{a}_C^*)^+$ (notation of §3) and, given any compact subset ω of $-Cl(\mathfrak{a}^*)^+$, we can choose a number c such that

$$|p(\Lambda)c_{ji}(s:\Lambda)| \leq c\,(1 + |\Lambda_I|)^{p_o}$$

for $\Lambda = \Lambda_R + i\Lambda_I$, $\Lambda_R \in \omega$ $\quad (p_o = \sum_{1\leq k\leq N} p_k)$.

Finally we prove the functional equation

$$(***) \qquad c_{ki}(ts:\Lambda) = c_{kj}(t:s\Lambda)c_{ji}(s:\Lambda)$$

for $s,t \in w(\mathfrak{a})$. First consider the case that $s = 1$. Choose a chain of chambers $C_{i_1},\ldots,C_{i_N}$ from C_i to C_j $(i_1 = i,\ i_N = j)$, as before. Put $c_p(\Lambda) = c_{i_{p+1},i_p}(1:\Lambda)$. One proves by induction, using Lemma 113, that

$$c_{ki_{N-p}}(t:\Lambda) = c_{kj}(t:\Lambda)c_{N-1}(\Lambda)\ldots c_{N-p}(\Lambda) \quad \text{for } 0 \leq p < N.$$

For $p = N-1$ this gives, in view of Lemma 117, $c_{ki}(t:\Lambda) = c_{kj}(t:\Lambda)c_{ji}(1:\Lambda)$. Taking adjoints we derive from the last equality that $c_{ik}(t^{-1}: -t\bar{\Lambda}) = c_{ij}(1: -\bar{\Lambda})c_{jk}(t^{-1}: -t\bar{\Lambda})$ (see Lemma 79), i.e. we have proved (***) for $t = 1$. In the general case, let m be the index such that $sC_i = C_m$. Then, by Lemma 114, $c_{ki}(ts:\Lambda) = c_{km}(t:s\Lambda)c_{mi}(s:\Lambda)$. And, applying what we just proved, $c_{mi}(s:\Lambda) = c_{mj}(1:s\Lambda)c_{ji}(s:\Lambda)$ and $c_{km}(t:s\Lambda)c_{mj}(1:s\Lambda) = c_{kj}(t:s\Lambda)$. This together gives the equality (***).

Theorem 8 is now proved.

Remark. As a consequence of Theorem 8 we have

$$c_{ij}(s^{-1}:s\Lambda)c_{ji}(s:\Lambda) = 1, \quad c_{ji}(s:\Lambda)c_{ij}(s^{-1}:s\Lambda) = 1,$$

so the $c_{ji}(s:\Lambda)$ are bijective "in general"; in particular $\dim \mathbb{L}_i = \dim \mathbb{L}_j$. By Lemma 79, $c_{ji}(s:\Lambda)^* = c_{ij}(s^{-1}: -s\bar{\Lambda})$, so $c_{ji}(s:\Lambda)$ is unitary if $Re(\Lambda) = 0$.

§5. Proof of Theorem 9.

Let (P,A) be a cuspidal pair with $P \in \mathcal{C}$. We use the notations of Chap.IV,§3. Let $\Phi = {}^{o}L^2(M/\Gamma_M, \sigma_M, \xi)$ with $\xi \in \mathcal{H}^o(P)$. Fix a real number R with $R > |\varrho|$. Define $\mathfrak{a}^*(R)$ and $\mathfrak{a}_c^*(R)$ as in Chap. IV,§3, and let $\mathcal{H}(R)$ denote the set of all bounded holomorphic functions h on $\mathfrak{a}_c^*(R)$ with values in Φ such that $\int_{\mathfrak{a}^*} |h(\lambda+i\mu)|^2 d\mu < \infty$ for every $\lambda \in \mathfrak{a}^*(R)$.

Let $\alpha_1,\ldots,\alpha_q$ be the roots in $\Sigma^o(P|A)$ and $\lambda_1,\ldots,\lambda_q$ the corresponding dual weights. Define the element H_o of $\mathfrak{a}$ by $\lambda_i(H_o) = 1$ $(1 \le i \le q)$. Given $x \in G$ and $N \in \mathbb{R}$, we define a function $v_{x,N}$ on A as follows. Put $H(x,N) = H(x) - NH_o$.

$$\begin{aligned} v_{x,N}(a) &= 1 \quad \text{if} \quad \lambda_i(\log a) > \lambda_i(H(x,N)) \quad \text{for} \quad 1 \le i \le q, \\ &= 0 \quad \text{otherwise.} \end{aligned}$$

For $x_o \in G$, $N \in \mathbb{R}$, $\varphi \in \Phi$ we define a "truncated Eisenstein series" by

$$E_{x_o,N}(\Lambda:\varphi:x) = \sum_{\Gamma/\Gamma\cap P} \varphi_\Lambda(x\gamma)\, v_{x_o,N}(a(x\gamma)) \qquad (\Lambda \in (\mathfrak{a}_c^*)^-,\ x \in G).$$

Here $\varphi_\Lambda(x) = \varphi(x)\, e^{(\Lambda-\varrho)(H(x))}$, as in Chap.II,§2; the series is of course absolutely convergent and its sum is continuous on $G \times (\mathfrak{a}_c^*)^-$, holomorphic in Λ.

Lemma 119. Fix $x_o \in G$, $N \in \mathbb{R}$, $\Lambda_o \in \mathfrak{a}_c^*$ and define $v_o(a) = v_{x_o,N}(a) e^{(\Lambda_o-\varrho)(\log a)}$ $(a \in A)$. Then the integral

$$\hat{v}_o(\Lambda) = \int_A v_o(a)\, e^{-(\Lambda-\varrho)(\log a)} da$$

exists and is equal to

$$b_o\, e^{-(\Lambda-\Lambda_o)(H(x_o,N))} \prod_{\alpha\in\Sigma^o} \langle \Lambda-\Lambda_o,\alpha\rangle^{-1}$$

provided $\mathrm{Re}\langle\Lambda-\Lambda_o,\alpha\rangle > 0$ for all $\alpha \in \Sigma^o(P|A)$. Here b_o is a positive constant given by $da = b_o \prod d\lambda_i$.

Proof. $\hat{v}_o(\Lambda) = b_o\, e^{(\Lambda_o-\Lambda)(H(x_o,N))} \int_{\lambda_i>0} e^{\Lambda_o-\Lambda} \prod d\lambda_i$.

Corollary. $\int_{\mathfrak{a}^*} |\hat{v}_o(\Lambda)|^2\, d\Lambda_I < \infty$ provided $\mathrm{Re}\langle\Lambda-\Lambda_o,\alpha\rangle > 0$ for all $\alpha \in \Sigma^o$.

Proof. $\int_{\mathfrak{a}^*} |\hat{v}_o(\Lambda+i\mu)|^2 \, d\mu = \int_A |v_o(a)|^2 e^{-2(\lambda-\varrho)(\log a)} da =$

$$b_o \int_{\lambda_i > \lambda_i(H(x_o,N))} e^{2(\lambda_o-\lambda)} \Pi \, d\lambda_i \quad \text{where} \quad \lambda_o = \mathrm{Re}(\Lambda_o) \; .$$

Let $D(R)$ denote the set of all $\Lambda \in \mathfrak{a}_C^*$ such that $\mathrm{Re}\langle\Lambda+\lambda,\alpha\rangle < 0$ for all $\alpha \in \Sigma^o$ and all $\lambda \in \mathrm{Cl}(\mathfrak{a}^*(R))$. Note that $\varrho \in \mathfrak{a}^*(R)$, so $D(R) \subset (\mathfrak{a}_C^*)^-$.

Lemma 120. Fix $\Lambda_o \in D(R)$, $\varphi \in \Phi$, $x_o \in G$, $N \in R$. Put

$$h(\Lambda) = b_o \, e^{-(\Lambda-\Lambda_o)(H(x_o,N))} \left(\prod_{\alpha\in\Sigma^o} \langle\Lambda-\Lambda_o,\alpha\rangle^{-1} \right) \varphi \qquad (\Lambda \in \mathfrak{a}_C^*(R)) \; .$$

Then $h \in \mathcal{H}(R)$ and $\check{E}_h(x) = E_{x_o,N}(\Lambda_o:\varphi:x)$ for almost all $x \in G$.

Proof. Define v_o as in Lemma 119. Then $h = \hat{v}_o \varphi \in \mathcal{H}(R)$ by the corollary of Lemma 119, the boundedness of h being obvious. By definition $\check{E}_h = E_{v_o\varphi}$. Choose a sequence of functions v_k^o in $C_c^\infty(A)$ such that $0 \le v_k^o \le v_{k+1}^o$ and $\lim v_k^o(a) = v_{x_o,N}(a)$ for all $a \in A$. Put $v_k(a) = v_k^o(a) e^{(\Lambda_o-\varrho)(\log a)}$. Then $\lim v_k(a) = v_o(a)$. Now $v_o \in L^2(A:\lambda)$ for $\lambda \in \mathfrak{a}^*(R)$ (notation of Chap.IV,§3), so that we have $\lim v_k = v_o$ in $L^2(A:\lambda)$. Put $f_k = v_k\varphi$ (i.e. $f_k(x) = v_k(a(x))\varphi(x)$). Then, by definition, $E_{v_o\varphi} = \lim E_{f_k}$ (in $L^2(G/\Gamma,\sigma)$). But, considering the series which defines E_{f_k} one sees immediately that $\lim E_{f_k}(x) = E_{x_o,N}(\Lambda_o:\varphi:x)$ for all x. Hence $E_{v_o\varphi}(x) = E_{x_o,N}(\Lambda_o:\varphi:x)$ for almost all x.

If $\chi \in \mathcal{H}_M$ we define as usual $\chi(z:\Lambda) = \chi(\mu(z:\Lambda))$ $(z \in \mathfrak{Z}, \Lambda \in \mathfrak{a}_C^*)$, μ being the canonical homomorphism $\mathfrak{Z} \to \mathfrak{Z}_M \mathfrak{A}$.

Lemma 121. Given $\chi \in \mathcal{H}_M$ and $p \in S(\mathfrak{a}_c)$, we can choose $z \in \mathfrak{Z}$ with the following property. There exists $q \neq o$ in $S(\mathfrak{a}_c)$ such that

$$\chi(z:\Lambda) = p(\Lambda)q(\Lambda) \qquad (\Lambda \in \mathfrak{a}_c^*) \; .$$

Proof. Extend $\mathfrak{a}$ to a Cartan subalgebra $\mathfrak{f}$ of $\mathfrak{g}$; $\mathfrak{f} = \mathfrak{a} + \mathfrak{b}$ with $\mathfrak{b} = \mathfrak{f} \cap \mathfrak{m}$. Let $I(\mathfrak{f}_C)$ (resp. $I(\mathfrak{b}_C)$) be the algebra of invariants of the Weyl group of $(\mathfrak{g},\mathfrak{f})$ (resp. $(\mathfrak{m},\mathfrak{b})$) in the symmetrical algebra $S(\mathfrak{f}_C)$ (resp. $S(\mathfrak{b}_C)$).

We have an isomorphism $S(\mathfrak{f}_C) \to S(\mathfrak{a}_C) \otimes S(\mathfrak{b}_C)$ which maps $I(\mathfrak{f}_C)$ into $S(\mathfrak{a}_C) \otimes I(\mathfrak{b}_C)$. By means of the canonical isomorphism $I(\mathfrak{b}_C) \to \mathfrak{Z}_M$ we derive from χ first a homomorphism $I(\mathfrak{b}_C) \to C$ and then a surjective homomorphism $S(\mathfrak{a}_C) \otimes I(\mathfrak{b}_C) \to S(\mathfrak{a}_C)$. Thus we have homomorphisms

$$\mathfrak{Z} \to I(\mathfrak{f}_C) \to S(\mathfrak{a}_C) \otimes I(\mathfrak{b}_C) \to S(\mathfrak{a}_C) .$$

Call the composition of these three homomorphisms φ_χ. Then $\chi(z:\Lambda) = (\varphi_\chi(z))(\Lambda)$. Since $S(\mathfrak{a}_C) \otimes I(\mathfrak{b}_C)$ is a finite module over $I(\mathfrak{f}_C)$, φ_χ makes $S(\mathfrak{a}_C)$ a finite $\mathfrak{Z}$-module. So, given an element $p \neq 0$ of $S(\mathfrak{a}_C)$, there are a natural number n and elements $z_1,\dots,z_n$ of $\mathfrak{Z}$ such that

$$p^n + \varphi_\chi(z_1)p^{n-1} + \dots + \varphi_\chi(z_n) = 0 , \quad \varphi_\chi(z_n) \neq 0 .$$

Then z_n has the required property.

<u>Corollary</u>. Given $\xi \in \mathcal{X}^0(P)$ and $p \in S(\mathfrak{a}_C)$, we can choose $z \in \mathfrak{Z}$ with the following property. For every $\chi \in \xi$ there exists $q_\chi \neq 0$ in $S(\mathfrak{a}_C)$ such that $\chi(z:\Lambda) = p(\Lambda)q_\chi(\Lambda)$ $(\Lambda \in \mathfrak{a}_C^*)$.

<u>Proof</u>. It is obviously enough to prove this under the assumption that p is invariant under $w(\mathfrak{a})$. Fix $\chi \in \xi$ and choose z as in Lemma 121 for this χ and p. Since ${}^s\mu(z) = \mu(z)$ for $z \in \mathfrak{Z}$, $s \in w(\mathfrak{a})$, we have ${}^s\chi(z:s\Lambda) = \chi(z:\Lambda)$. So, if $\chi(z:\Lambda) = p(\Lambda)q(\Lambda)$, then ${}^s\chi(z:\Lambda) = \chi(z:s^{-1}\Lambda) = p(\Lambda)q(s^{-1}\Lambda)$.

As a consequence of Lemma 118 it is possible to choose p in $S(\mathfrak{a}_C)$, $p \neq 0$, with the following properties. Firstly the functions $p(\Lambda)c(s:\Lambda)$ $(s \in w(\mathfrak{a}))$ are holomorphic for $\Lambda \in -Cl(\mathfrak{a}_C^*)^+$. Moreover, given any compact subset ω of $-Cl(\mathfrak{a}^*)^+$, there is a number $c(\omega)$ such that $|p(\Lambda)c(s:\Lambda)| \leq c(\omega)\ (1 + |\Lambda_I|)^d$ for $\Lambda_R \in \omega$ (d is the degree of p). Fix such a p and choose z in $\mathfrak{Z}$ as in the corollary of Lemma 121 corresponding to this p.

Fix $\varphi_1,\varphi_2 \in \Phi$ and $\Lambda_1,\Lambda_2 \in D(R)$. Put

$$h_i(\Lambda) = b_0\, e^{-(\Lambda-\Lambda_i)(H(x_0,N))} \Big(\prod_{\beta\in\Sigma^0} \langle \Lambda-\Lambda_i,\beta\rangle^{-1}\Big)\varphi_i \qquad (\Lambda \in \mathfrak{a}_C^*(R))$$

for $i = 1,2$ and some fixed x_o, N. Put $E'(\Lambda_i:\varphi_i) = E_{x_o,N}(\Lambda_i:\varphi_i)$. Then $E'(\Lambda_i:\varphi_i) = \check{E}_{h_i}$ (Lemma 120). Fix $\alpha \in I_c^\infty(G)$ and put $E''(\Lambda_i:\varphi_i) = z\alpha * E'(\Lambda_i:\varphi_i)$. Then $E''(\Lambda_i:\varphi_i) = \check{E}_{h_i'}$ where h_i' is defined by

$$h_i'(\Lambda) = \hat{\pi}(\alpha:\Lambda)\xi(z:\Lambda)h_i(\Lambda) \qquad (\Lambda \in \mathfrak{a}_C^*(R))$$

(see Lemmas 75 and 74). Recall that $\hat{\pi}(\alpha)$ is the Fourier transform of a C^∞ function with compact support on A (with values in End(Φ)). So it follows from our assumptions that $c(s:\Lambda)\hat{\pi}(\alpha:\Lambda)\xi(z:\Lambda)$ is holomorphic for $\Lambda \in -\mathrm{Cl}(\mathfrak{a}_C^*)^+$ and, given any compact subset ω of $-\mathrm{Cl}(\mathfrak{a}^*)^+$ and any $r \geq 0$, we can choose $c_o \geq 0$ such that

$$|c(s:\Lambda)\hat{\pi}(\alpha:\Lambda)\xi(z:\Lambda)|\ (1 + |\Lambda_I|)^r \leq c_o \quad \text{if} \quad \Lambda_R \in \omega .$$

So both $h_1'(\Lambda)$ and $c(s:\Lambda)h_2'(\Lambda)$ are rapidly decreasing functions of Λ_I. The scalar-product formula gives

$$((E''(\Lambda_1:\varphi_1),\ E''(\Lambda_2:\varphi_2)))_{G/\Gamma} = \sum_{s\in W(\mathfrak{a})} \int_{\mathfrak{a}^*} (h_1'(-s\bar{\Lambda}),\ c(s:\Lambda)h_2'(\Lambda))_\Phi\, d\Lambda_I$$

where $\Lambda_R \in \mathfrak{a}^*(R) \cap (\mathfrak{a}^*)^-$. Put, for $\Lambda \in \mathfrak{a}_C^*(R) \cap -\mathrm{Cl}(\mathfrak{a}_C^*)^+$,

$$J(\Lambda) = \sum_{s\in W(\mathfrak{a})} \int_{\mathfrak{a}^*} (h_1'(-s(\bar{\Lambda}-i\mu)), c(s:\Lambda+i\mu)h_2'(\Lambda+i\mu))_\Phi\, d\mu .$$

One sees immediately that J is continuous on $\mathfrak{a}_C^*(R) \cap -\mathrm{Cl}(\mathfrak{a}_C^*)^+$ and holomorphic on $\mathfrak{a}_C^*(R) \cap -(\mathfrak{a}_C^*)^+$; since it is independant of Λ_I, it is therefore constant. Hence

$$((E''(\Lambda_1:\varphi_1), E''(\Lambda_2:\varphi_2)))_{G/\Gamma} = \sum_{s\in W(\mathfrak{a})} \int_{\mathfrak{a}^*} (h_1'(is\mu), c(s:i\mu)h_2'(i\mu))_\Phi\, d\mu .$$

Call the right hand side of this formula $J(\bar{\Lambda}_1, \Lambda_2)$. Since

$$\|h_1'(is\mu)\| \leq c(1+|\mu|)^{-r}\|h_1(is\mu)\|, \quad \|c(s:i\mu)h_2'(i\mu)\| \leq c(1+|\mu|)^{-r}\|h_2(i\mu)\|$$

(where r is arbitrary, $c^\cdot$ depends on r, α, z, but not on Λ_1, Λ_2), and

$$\|h_i(\Lambda)\| \leq b_o\, e^{\mathrm{Re}\,\Lambda_i(H(x_o,N))} \Big(\prod_{\beta\in\Sigma^o} |\mathrm{Re}\langle\Lambda_i,\beta\rangle|^{-1}\Big)\|\varphi_i\| \quad \text{for} \quad \mathrm{Re}(\Lambda) = o,\ \Lambda_i \in -(\mathfrak{a}_C^*)^+,$$

the integrals in the definition of $J(\bar{\Lambda}_1,\Lambda_2)$ converge uniformly if Λ_1,Λ_2 stay in compact subsets of $-(\mathfrak{a}_C^*)^+$. So J is holomorphic on $-(\mathfrak{a}_C^*)^+ \times -(\mathfrak{a}_C^*)^+$. Hence, for any $\varphi \in \Phi$, the mapping $\Lambda \longmapsto E''(\Lambda:\varphi)$ extends to a holomorphic mapping of $-(\mathfrak{a}_C^*)^+$ into $\mathrm{Cl}\ E_{P|A}(\sigma,\xi)$ (cf. the proof of Lemma 87). Moreover,

$$\|E''(\Lambda:\varphi)\| = |J(\bar{\Lambda},\Lambda)|^{1/2} \leq c\, e^{\mathrm{Re}\ \Lambda(H(x_o,N))} (\Pi |\mathrm{Re}\langle\Lambda,\beta\rangle|^{-1}) \|\varphi\| .$$

We have proved:

<u>Lemma 122</u>. Fix $z \in \mathfrak{Z}$ as above. Then for any $\alpha \in I_c^\infty(G)$ we can choose a constant c with the following property. Fix $x_o \in G$, $N \in R$, $\varphi \in \Phi$, and put

$$E''(\Lambda:\varphi) = z\alpha * E_{x_o,N}(\Lambda:\varphi)$$

for $\Lambda \in D(R)$. Then $E''(\varphi)$ extends to a holomorphic mapping of $-(\mathfrak{a}_C^*)^+$ into $\mathrm{Cl}\ E_{P|A}(\sigma,\xi)$ and $\|E''(\Lambda:\varphi)\|_{G/\Gamma} \leq c\|\varphi\|\, e^{\Lambda_R(H(x_o,N))} \prod_{\beta\in\Sigma^o} |\langle\Lambda_R,\beta\rangle|^{-1}$ for $\Lambda \in -(\mathfrak{a}_C^*)^+$, $\varphi \in \Phi$, where c is independant of Λ, φ, x_o, N.

Now we consider $E''(\Lambda:\varphi)$ as a distribution, i.e. we define

$$E''(\Lambda:\varphi:\beta) = \int_G \beta(x) E''(\Lambda:\varphi:x)\,dx$$

for $\Lambda \in -(\mathfrak{a}_C^*)^+$, $\beta \in C_c^\infty(G)$.

<u>Lemma 123</u>. Let ω be a compact subset of $-(\mathfrak{a}_C^*)^+$. Then for any $\varphi \in \Phi$ the distributions $E''(\Lambda:\varphi)$ are uniformly continuous on $C_c^\infty(G)$ for $\Lambda \in \omega$.

<u>Proof</u>. Fix a compact subset Ω of G. We have

$$|E''(\Lambda:\varphi:\beta)| \leq \int_G |\beta(x)|\ |E''(\Lambda:\varphi:x)|\,dx = \int_{G/\Gamma} \sum_{\Gamma} \leq$$

$$\leq c_1\|\beta\|_\infty \int_{G/\Gamma} |E''(\Lambda:\varphi:x)|\,dx \leq c_1\|\beta\|_\infty \|E''(\Lambda:\varphi)\|_{G/\Gamma} \leq c\|\beta\|_\infty$$

for $\Lambda \in \omega$, $\beta \in C_\Omega^\infty(G)$.

<u>Corollary</u>. $E''(\Lambda:\varphi:\beta)$ is a continuous function of (Λ,β) on $-(\mathfrak{a}_C^*)^+ \times C_c^\infty(G)$ which is holomorphic in Λ.

Let (P_o,A_o) be a mincuspidal pair of G dominated by (P,A) and $\mathfrak{S}$ a Siegel domain with respect to (P_o,A_o).

<u>Lemma 124</u>. Let Ω be a compact subset of G. There exists a real number N_o such that

$$E(\Lambda:\varphi:y_1x_oy_2) = E_{x_o,N}(\Lambda:\varphi:y_1x_oy_2)$$

for $y_1,y_2 \in \Omega$, $x_o \in \mathfrak{S}$, $\varphi \in \Phi$, $\Lambda \in (\mathfrak{a}_C^*)^-$, $N \geq N_o$.

<u>Proof</u>. There exists N_o such that $\lambda(H(y_1xy_2\gamma) - H(x)) > -N_o$ for $y_1,y_2 \in \Omega$, $x \in \mathfrak{S}$, $\gamma \in \Gamma$ and any simple dual weight λ of (P,A) (Lemma 22). If $N \geq N_o$, then $v_{x_o,N}(a(y_1x_oy_2\gamma)) = 1$ for $y_1,y_2 \in \Omega$, $x_o \in \mathfrak{S}$, $\gamma \in \Gamma$.

<u>Corollary 1</u>. Let Ω be a compact subset of G. There exists a real number N such that

$$E(\Lambda:\varphi:x) = E_{x_o,N}(\Lambda:\varphi:x)$$

for $x,x_o \in \Omega$, $\Lambda \in (\mathfrak{a}_C^*)^-$, $\varphi \in \Phi$.

<u>Corollary 2</u>. Let Ω be a compact subset of G. There exists a real number N with the following property. Let $\alpha \in C_\Omega(G)$ and $x_o \in \Omega$. Then

$$\alpha * E(\Lambda:\varphi) = \alpha * E_{x_o,N}(\Lambda:\varphi) \quad \text{on } \Omega$$

for $\Lambda \in (\mathfrak{a}_C^*)^-$ and $\varphi \in \Phi$.

Fix $z \in \mathfrak{Z}$ as before. Put $D = -(\mathfrak{a}_C^*)^+$ and let D' be the set of all $\Lambda \in D$ such that $\det(\xi(z:\Lambda)) \neq 0$ (note that this determinant is not identically zero).

<u>Lemma 125</u>. For any $\varphi \in \Phi$, $E(\Lambda:\varphi:x)$ extends to a C^∞ function on $D' \times G$ which, for any fixed $x \in G$, is meromorphic for $\Lambda \in D$ and holomorphic for $\Lambda \in D'$. Moreover, given any compact subset ω of D', we can choose c and N such that

$$|E(\Lambda:\varphi:x)| \leq c\|\varphi\| \|x\|^N.$$

for all $\Lambda \in \omega$, $\varphi \in \Phi$, $x \in G$.

<u>Proof</u>. Fix $\alpha_o \in I_c^{\infty}(G)$. Put $E_o(\Lambda:\varphi) = z\alpha_o * E(\Lambda:\varphi)$ $(\Lambda \in (\mathfrak{a}_c^*)^-)$. Fix an open relatively compact set Ω in G such that $\Omega = \Omega^{-1} \supset \mathrm{Supp}(\alpha_o)$. Put $\Omega_o = \Omega \cup \Omega\Omega$. By Corollary 2 of Lemma 124 we can choose $N_o \in R$ such that

$$\alpha * E(\Lambda:\varphi) = \alpha * E_{x_o,N_o}(\Lambda:\varphi) \quad \text{on} \quad \Omega_o$$

for $x_o \in \Omega_o$, $\alpha \in C_c(\Omega_o)$, $\varphi \in \Phi$, $\Lambda \in (\mathfrak{a}_c^*)^-$. Fix $x_o \in \Omega_o$ and put

$$E''(\Lambda:\varphi) = z\alpha_o * E_{x_o,N_o}(\Lambda:\varphi) \qquad (\varphi \in \Phi,\ \Lambda \in D(R)) .$$

By Lemma 122, $E''(\varphi)$ extends to a holomorphic mapping of D into $\mathrm{Cl}\ E_{P|A}(\sigma,\xi)$. We have $E''(\Lambda:\varphi) = E_o(\Lambda:\varphi)$ on Ω_o if $\Lambda \in D(R)$. So $E''(\Lambda:\varphi)$ and $E_o(\Lambda:\varphi)$ are equal as distributions on Ω_o , i.e.

$$E_o(\Lambda:\varphi:\beta) = E''(\Lambda:\varphi:\beta) \quad \text{for} \quad \Lambda \in D(R),\ \beta \in C_c^{\infty}(\Omega_o) .$$

In particular $E_o(\Lambda:\varphi:\alpha_o' * \beta) = E''(\Lambda:\varphi:\alpha_o' * \beta)$ if $\Lambda \in D(R)$, $\beta \in C_c^{\infty}(\Omega)$, $\alpha_o'(x) = \alpha_o(x^{-1})$; in other words $\alpha_o * E_o(\Lambda:\varphi) = \alpha_o * E''(\Lambda:\varphi)$ as distributions on Ω if $\Lambda \in D(R)$. Now $\alpha_o * E_o(\Lambda:\varphi) = z\alpha_o * \alpha_o * E(\Lambda:\varphi) = z\alpha_o * E(\Lambda:\pi(\alpha_o:\Lambda)\varphi) = E_o(\Lambda:\pi(\alpha_o:\Lambda)\varphi)$; on the other hand $E_o(\Lambda:\pi(\alpha_o:\Lambda)\varphi) = E''(\Lambda:\pi(\alpha_o:\Lambda)\varphi)$ as distributions on Ω_o $(\Lambda \in D(R))$. Hence, for $\Lambda \in D(R)$,

$$E''(\Lambda:\pi(\alpha_o:\Lambda)\varphi) = \alpha_o * E''(\Lambda:\varphi) \quad \text{as distributions on } \Omega.$$

Then this holds also for $\Lambda \in D$, since $E''(\Lambda:\varphi:\beta)$ is a holomorphic function of Λ for $\Lambda \in D$. But $\alpha_o * E''(\Lambda:\varphi)$ is a C^{∞} function on G ; so $E''(\Lambda:\pi(\alpha_o:\Lambda)\varphi)$ is a C^{∞} function on Ω $(\Lambda \in D)$. Moreover, if $g \in \mathfrak{g}$,

$$E''(\Lambda:\pi(\alpha_o:\Lambda)\varphi : g;x) = \int_G \alpha_o(g;xy^{-1})E''(\Lambda:\varphi:y)dy ,$$

and this is a continuous function of (Λ,x) on $D \times \Omega$. Recalling that $E_o(\Lambda:\pi(\alpha_o:\Lambda)\varphi:x) = E''(\Lambda:\pi(\alpha_o:\Lambda)\varphi:x)$ for $\Lambda \in D(R)$, $x \in \Omega$, we see that, for any $x \in \Omega$, $E_o(\Lambda:\pi(\alpha_o:\Lambda)\varphi:x)$ extends to a holomorphic function of Λ on D and $E_o(\Lambda:\pi(\alpha_o:\Lambda)\varphi:x)$ is a C^{∞} function of (Λ,x) on $D \times \Omega$. Since we can take Ω arbitrarily large we have in fact a C^{∞} function on $D \times G$. Now $E_o(\Lambda:\varphi) = E(\Lambda:\pi(\alpha_o:\Lambda)\xi(z:\Lambda)\varphi)$. Given

$\Lambda_o \in D$, we can choose α_o such that $\pi(\alpha_o:\Lambda_o) = 1$. Then, for Λ in a neighbourhood of Λ_o , $E(\Lambda:\xi(z:\Lambda)\varphi) = E_o(\Lambda:\pi(\alpha_o:\Lambda)\pi(\alpha_o:\Lambda)^{-2}\varphi)$. So it follows from what we proved above that $E(\Lambda:\xi(z:\Lambda)\varphi:x)$ is holomorphic for $\Lambda \in D$ and C^∞ on $D \times G$, and $E(\Lambda:\varphi:x)$ is holomorphic for $\Lambda \in D'$, meromorphic for $\Lambda \in D$ and C^∞ on $D' \times G$.

In order to prove the last assertion of Lemma 125 we fix a mincuspidal pair (P_o,A_o) dominated by (P,A) and a Siegel domain $\mathfrak{S}$ with respect to (P_o,A_o) and a finite subset Ξ of $G_{\mathbb{Q}}$ such that $G = \mathfrak{S}\Xi\Gamma$. Fix α_o , Ω and Ω_o as in the first part of the proof. Choose Ω such that it contains also Ξ . Define as before $E_o(\Lambda:\varphi) = z\alpha_o * E(\Lambda:\varphi)$ $(\Lambda \in (\mathfrak{a}_C^*)^-)$. By Lemma 124 we can choose $N \in R$ such that

$$E(\Lambda:\varphi:x) = E_{x_o,N}(\Lambda:\varphi:x)$$

for $\Lambda \in (\mathfrak{a}_C^*)^-$, $\varphi \in \Phi$, $x \in \Omega_o x_o \Omega_o$, $x_o \in \mathfrak{S}$. Put

$$E''_{x_o}(\Lambda:\varphi) = z\alpha_o * E_{x_o,N}(\Lambda:\varphi) \qquad (\Lambda \in D(R),\ x_o \in \mathfrak{S})$$

and

$$E'''_{x_o}(\Lambda:\varphi) = \alpha_o * E''_{x_o}(\Lambda:\varphi) .$$

Then $E'''_{x_o}(\Lambda:\varphi) = \alpha_o * E_o(\Lambda:\varphi)$ on $x_o\Omega_o$ $(x_o \in \mathfrak{S},\ \Lambda \in D(R))$.

$$|E'''_{x_o}(\Lambda:\varphi:x)| \leq \int_G |\alpha_o(xy^{-1})|\ |E''_{x_o}(\Lambda:\varphi:y)|dy = \int_{G/\Gamma} \sum_{\Gamma} \leq$$

$$c_1\|x\|^{N_1}\ \|E''_{x_o}(\Lambda:\varphi)\|_{G/\Gamma} \leq c_2\|x\|^{N_1}\ \|\varphi\|\ e^{\Lambda_R(H(x_o,N))} \prod_{\beta\in\Sigma^o(P|A)} |\langle\Lambda_R,\beta\rangle|^{-1}$$

for $x_o \in \mathfrak{S}$, $\Lambda \in D$, $\varphi \in \Phi$, $x \in G$ (Lemma 122). Let ω be a compact subset of D . If $x_o \in \mathfrak{S}$, $\Lambda \in \omega$, $\varphi \in \Phi$, $x \in x_o\Xi$, then

$$|E'''_{x_o}(\Lambda:\varphi:x)| \leq c_3\|x\|^{N_1}\ \|\varphi\|\ \|x_o\|^{N_2} \leq c_4\|x\|^{N_3}\ \|\varphi\| . \quad \text{So}$$

$$|E(\Lambda:\pi(\alpha_o:\Lambda)^2\xi(z:\Lambda)\varphi:x)| \leq c_4\|x\|^{N_3}\ \|\varphi\|$$

for $\Lambda \in \omega$, $\varphi \in \Phi$, $x \in \mathfrak{S}\Xi$. Hence

$$|E(\Lambda:\pi(\alpha_o:\Lambda)^2\xi(z:\Lambda)\varphi:x)| \leq c_5\|x\|^{N_3}\ \|\varphi\| \quad \text{for } \Lambda \in \omega,\ \varphi \in \Phi,\ x \in G .$$

Given $\Lambda_o \in D'$, there are a neighbourhood ω_o of Λ_o in D' and numbers c, M such that

$$|E(\Lambda:\varphi:x)| \leq c\|x\|^M\|\varphi\| \quad \text{for } \Lambda \in \omega_o,\ \varphi \in \Phi,\ x \in G .$$

Lemma 125 is now completely proved.

Corollary. If $f = E(\Lambda:\varphi)$ and if (P',A') is a cuspidal pair associated to (P,A), then

$$f_{P'}(x) = \sum_{s\in W(\mathfrak{a},\mathfrak{a}')} (c_{P'|P}(s:\Lambda)\varphi)(x)\ e^{(s\Lambda-\varrho')(H'(x))} \quad (\Lambda \in D') .$$

Let $P_1,\dots,P_r$ be the cuspidal groups in $\mathcal{C}$ which have A as a split component. They correspond to the distinct chambers $C_1,\dots,C_r$ of $\mathfrak{a}$. We use now the notations of §2 with these P_i. Define $\underline{\underline{L}} = \prod_{1\leq i\leq r} \underline{\underline{L}}_i$; define the linear transformation $\underline{\underline{c}}(s:\Lambda) : \underline{\underline{L}} \to \underline{\underline{L}}$ by

$$(\underline{\underline{c}}(s:\Lambda)\varphi)_j = c_{ji}(s:\Lambda)\varphi \quad \text{if} \quad \varphi \in \underline{\underline{L}}_i \qquad (\Lambda \in \mathfrak{a}_C^*) .$$

Put $\underline{\underline{\Phi}} = \underline{\underline{L}}^W$, where $W = W(\mathfrak{a})$. For any $\Lambda \in \mathfrak{a}_C^*$ let $L(\Lambda)$ denote the set of all functions f in $\mathcal{A}_q(G/\Gamma,\sigma)$ for which there exists an element $h = (h_s)$ of $\underline{\underline{\Phi}}$ such that

1) $f_P = 0$ if rank $P = q$, $P \notin \mathcal{C}$;
2) $f_{P_{ik}}(x) = \sum_{s\in W} (h_s)_{ik}(x)\ e^{Y_{ik}(s\Lambda-\varrho_i)(H_{ik}(x))}$ (all i,k) .

Let $\underline{\underline{\Phi}}(\Lambda)$ denote the set of all elements h of $\underline{\underline{\Phi}}$ for which there exists $f \in \mathcal{A}_q(G/\Gamma,\sigma)$ with these properties. We have a surjective linear mapping $h \longmapsto f(h:\Lambda)$ of $\underline{\underline{\Phi}}(\Lambda)$ on $L(\Lambda)$ (this mapping is not necessarily bijective).

Lemma 126. Let Λ_o be a point in $\mathfrak{a}_C^*$ such that $\underline{\underline{c}}(s:\Lambda)$ $(s \in W)$ are all holomorphic at Λ_o. For any $\varphi \in \underline{\underline{L}}$ define $h(\varphi) \in \underline{\underline{\Phi}}$ by

$$h_s(\varphi) = \underline{\underline{c}}(s:\Lambda_o)\varphi \quad (s \in W) .$$

Then $h(\varphi) \in \underline{\underline{\Phi}}(\Lambda_o)$.

Proof. For $1 \leq j \leq r$, let D_j denote the set of all $\Lambda \in \mathfrak{a}_C^*$ such that

$\mathrm{Re}\langle\Lambda,\alpha\rangle < 0$ for all $\alpha \in \Sigma(P_j|A)$ and let D_j' denote the set of all $\Lambda_1 \in D_j$ such that $\underline{\underline{E}}(P_j|A:\Lambda:\varphi:x)$ is holomorphic at Λ_1 for any $\varphi \in \underline{\underline{L}}_j$ and any $x \in G$. By Lemma 125, D_j' contains all points of D_j where a certain polynomial does not vanish.

It is enough to prove Lemma 126 for a function φ in $\underline{\underline{L}}_i$. Choose j such that $\Lambda_o \in \mathrm{Cl}\, D_j$. Fix a neighbourhood D^o of Λ_o in $\mathfrak{a}_c^*$ such that the $\underline{\underline{c}}(s:\Lambda)$ are all holomorphic on D^o. Define

$$f(\Lambda:x) = \underline{\underline{E}}(P_j|A:\Lambda:c_{ji}(1:\Lambda)\varphi:x) \qquad (\Lambda \in D^o \cap D_j') .$$

Then, for $\Lambda \in D^o \cap D_j'$,

$$f_{P_{kl}}(\Lambda:x) = \sum_{s\in w} (c_{kj}(s:\Lambda)c_{ji}(1:\Lambda)\varphi)_{kl}(x)e^{y_{kl}(s\Lambda-\rho_k)(H_{kl}(x))}$$

$$= \sum_{s\in w} (c_{ki}(s:\Lambda)\varphi)_{kl}(x)e^{y_{kl}(s\Lambda-\rho_k)(H_{kl}(x))} .$$

This shows that the element $h(\Lambda)$ of $\underline{\underline{\Phi}}$ defined by $h_s(\Lambda) = \underline{\underline{c}}(s:\Lambda)\varphi$ $(s \in w)$ belongs to $\underline{\underline{\Phi}}(\Lambda)$ for $\Lambda \in D^o \cap D_j'$. Now make Λ tend to Λ_o; then $h(\Lambda)$ tends to $h(\varphi)$ in $\underline{\underline{\Phi}}$. So $h(\varphi) \in \underline{\underline{\Phi}}(\Lambda_o)$ by Corollary 1 of Theorem 6'.

Let $D(A)$ be the set of all points Λ_o of $\mathfrak{a}_c^*$ such that $\underline{\underline{c}}(s:\Lambda)$ $(s \in w)$ are all holomorphic at Λ_o. Then $D(A)$ is an open dense connected subset of $\mathfrak{a}_c^*$. For any $\varphi \in \underline{\underline{L}}$, $\Lambda \in D(A)$, define

$$\underline{\underline{E}}(\Lambda:\varphi) = f(h(\Lambda:\varphi):\Lambda)$$

where $h_s(\Lambda:\varphi) = \underline{\underline{c}}(s:\Lambda)\varphi$ $(s \in w)$.

<u>Lemma 127</u>. $\underline{\underline{E}}(\varphi)$ is a C^∞ function on $D(A) \times G$; for any fixed $x \in G$, $\underline{\underline{E}}(\Lambda:\varphi:x)$ is a meromorphic function of Λ on $\mathfrak{a}_c^*$ which is holomorphic on $D(A)$. Given a compact subset ω of $D(A)$, we can choose c and N such that

$$|\underline{\underline{E}}(\Lambda:\varphi:x)| \leq c\|x\|^N \|\varphi\|_{\underline{\underline{L}}}$$

for $\Lambda \in \omega$, $\varphi \in \underline{\underline{L}}$, $x \in G$.

<u>Proof</u>. This follows immediately from Theorem 6' and its corollaries.

One sees from the proof of Lemma 126 that $\underline{\underline{E}}(\Lambda:\varphi) = \underline{\underline{E}}(P_i|A:\Lambda:\varphi)$ if $\varphi \in \underline{\underline{L}}_i$ and $\Lambda \in D(A) \cap \mathrm{Cl}\, D_i$. Thus we have

$$\underline{\underline{E}}(\Lambda:\varphi) = \sum_{1 \leq i \leq r} \underline{\underline{E}}(P_i|A:\Lambda:\varphi_i) \qquad (\varphi = (\varphi_i) \in \underline{\underline{L}}).$$

Theorem 9 is now proved.

The functional equation for $\underline{\underline{E}}(\Lambda:\varphi)$ reads

$$\underline{\underline{E}}(s\Lambda:\underline{\underline{c}}(s:\Lambda)\varphi) = r\ \underline{\underline{E}}(\Lambda:\varphi) \qquad (s \in w,\ \varphi \in \underline{\underline{L}},\ \Lambda \in \mathfrak{a}_C^*).$$

Offsetdruck: Julius Beltz, Weinheim/Bergstr.

Lecture Notes in Mathematics

Bisher erschienen/Already published

Vol. 1: J. Wermer, Seminar über Funktionen-Algebren. IV, 30 Seiten. 1964. DM 3,80 / $ 0.95

Vol. 2: A. Borel, Cohomologie des espaces localement compacts d'après J. Leray. IV, 93 pages. 1964. DM 9,– / $ 2.25

Vol. 3: J. F. Adams, Stable Homotopy Theory. 2nd. revised edition. IV, 78 pages. 1966. DM 7,80 / $ 1.95

Vol. 4: M. Arkowitz and C. R. Curjel, Groups of Homotopy Classes. 2nd. revised edition. IV, 36 pages. 1967. DM 4,80 / $ 1.20

Vol. 5: J.-P. Serre, Cohomologie Galoisienne. Troisième édition. VIII, 214 pages. 1965. DM 18,– / $ 4.50

Vol. 6: H. Hermes, Eine Termlogik mit Auswahloperator. IV, 42 Seiten. 1965. DM 5,80 / $ 1.45

Vol. 7: Ph. Tondeur, Introduction to Lie Groups and Transformation Groups. VIII, 176 pages. 1965. DM 13,50 / $ 3.40

Vol. 8: G. Fichera, Linear Elliptic Differential Systems and Eigenvalue Problems. IV, 176 pages. 1965. DM 13.50 / $ 3.40

Vol. 9: P. L. Ivănescu, Pseudo-Boolean Programming and Applications. IV, 50 pages. 1965. DM 4,80 / $ 1.20

Vol. 10: H. Lüneburg, Die Suzukigruppen und ihre Geometrien. VI, 111 Seiten. 1965. DM 8,– / $ 2.00

Vol. 11: J.-P. Serre, Algèbre Locale. Multiplicités. Rédigé par P. Gabriel. Seconde édition. VIII, 192 pages. 1965. DM 12,– / $ 3.00

Vol. 12: A. Dold, Halbexakte Homotopiefunktoren. II, 157 Seiten. 1966. DM 12,– / $ 3.00

Vol. 13: E. Thomas, Seminar on Fiber Spaces. IV, 45 pages. 1966. DM 4,80 / $ 1.20

Vol. 14: H. Werner, Vorlesung über Approximationstheorie. IV, 184 Seiten und 12 Seiten Anhang. 1966. DM 14,– / $ 3.50

Vol. 15: F. Oort, Commutative Group Schemes. VI, 133 pages. 1966. DM 9,80 / $ 2.45

Vol. 16: J. Pfanzagl and W. Pierlo, Compact Systems of Sets. IV, 48 pages. 1966. DM 5,80 / $ 1.45

Vol. 17: C. Müller, Spherical Harmonics. IV, 46 pages. 1966. DM 5,– / $ 1.25

Vol. 18: H.-B. Brinkmann und D. Puppe, Kategorien und Funktoren. XII, 107 Seiten. 1966. DM 8,– / $ 2.00

Vol. 19: G. Stolzenberg, Volumes, Limits and Extensions of Analytic Varieties. IV, 45 pages. 1966. DM 5,40 / $ 1.35

Vol. 20: R. Hartshorne, Residues and Duality. VIII, 423 pages. 1966. DM 20,– / $ 5.00

Vol. 21: Seminar on Complex Multiplication. By A. Borel, S. Chowla, C. S. Herz, K. Iwasawa, J.-P. Serre. IV, 102 pages. 1966. DM 8,– / $ 2.00

Vol. 22: H. Bauer, Harmonische Räume und ihre Potentialtheorie. IV, 175 Seiten. 1966. DM 14,– / $ 3.50

Vol. 23: P. L. Ivănescu and S. Rudeanu, Pseudo-Boolean Methods for Bivalent Programming. 120 pages. 1966. DM 10,– / $ 2.50

Vol. 24: J. Lambek, Completions of Categories. IV, 69 pages. 1966. DM 6,80 / $ 1.70

Vol. 25: R. Narasimhan, Introduction to the Theory of Analytic Spaces. IV, 143 pages. 1966. DM 10,– / $ 2.50

Vol. 26: P.-A. Meyer, Processus de Markov. IV, 190 pages. 1967. DM 15,– / $ 3.75

Vol. 27: H. P. Künzi und S. T. Tan, Lineare Optimierung großer Systeme. VI, 121 Seiten. 1966. DM 12,– / $ 3.00

Vol. 28: P. E. Conner and E. E. Floyd, The Relation of Cobordism to K-Theories. VIII, 112 pages. 1966. DM 9,80 / $ 2.45

Vol. 29: K. Chandrasekharan, Einführung in die Analytische Zahlentheorie. VI, 199 Seiten. 1966. DM 16,80 / $ 4.20

Vol. 30: A. Frölicher and W. Bucher, Calculus in Vector Spaces without Norm. X, 146 pages. 1966. DM 12,– / $ 3.00

Bitte wenden / Continued

Vol. 31: Symposium on Probability Methods in Analysis Chairman: D.A Kappos. IV, 329 pages 1967. DM 20,– / $ 5.00

Vol. 32: M. André, Méthode Simpliciale en Algèbre Homologique et Algèbre Commutative. IV, 122 pages. 1967. DM 12,– / $ 3.00

Vol. 33: G. I. Targonski, Seminar on Functional Operators and Equations. IV, 110 pages. 1967. DM 10,– / $ 2.50

Vol. 34: G. E. Bredon, Equivariant Cohomology Theories. VI, 64 pages. 1967. DM 6,80 / $ 1.70

Vol. 35: N. P. Bhatia and G. P. Szegö, Dynamical Systems: Stability Theory and Applications. VI, 416 pages. 1967. DM 24,– / $ 6.00

Vol. 36: A. Borel, Topics in the Homology Theory of Fibre Bundles. VI, 95 pages. 1967. DM 9,– / $ 2.25

Vol. 37: R. B. Jensen, Modelle der Mengenlehre. X, 176 Seiten. 1967. DM 14,– / $ 3.50

Vol. 38: R. Berger, R. Kiehl, E. Kunz und H.-J. Nastold. Differentialrechnung in der analytischen Geometrie. IV, 134 Seiten. 1967. DM 12,– / $ 3.00

Vol. 39: Séminaire de Probabilités I. II, 189 pages. 1967. DM 14,– / $ 3.50

Vol. 40: J. Tits, Tabellen zu den einfachen Lie Gruppen und ihren Darstellungen. VI, 53 Seiten. 1967. DM 6,80 / $ 1.70

Vol. 41: A. Grothendieck, Local Cohomology. VI, 106 pages. 1967. DM 10,– / $ 2.50

Vol. 42: J. F. Berglund and K. H. Hofmann, Compact Semitopological Semigroups and Weakly Almost Periodic Functions. VI, 160 pages. 1967. DM 12,– / $ 3.00

Vol. 43: D. G. Quillen, Homotopical Algebra. VI, 157 pages. 1967. DM 14,– / $ 3.50

Vol. 44: K. Urbanik, Lectures on Prediction Theory. IV, 50 pages. 1967. DM 5,80 / $ 1.45

Vol. 45: A. Wilansky, Topics in Functional Analysis. VI, 102 pages. 1967. DM 9,60 / $ 2.40

Vol. 46: P. E. Conner, Seminar on Periodic Maps. IV, 116 pages. 1967. DM 10,60 / $ 2.65

Vol. 47: Reports of the Midwest Category Seminar. IV, 181 pages. 1967. DM 14,80 / $ 3.70

Vol. 48: G. de Rham, S. Maumary and M. A. Kervaire, Torsion et Type Simple d'Homotopie. IV, 101 pages. 1967 DM 9,60 / $ 2.40

Vol. 49: C. Faith, Lectures on Injective Modules and Quotient Rings. XVI, 140 pages. 1967. DM 12,80 / $ 3.20

Vol. 50: L. Zalcman, Analytic Capacity and Rational Approximation. VI. 155 pages. 1968. DM 13,20/$ 3.40

Vol. 51: Séminaire de Probabilités II. IV, 199 pages. 1968. DM 14,–/$ 3.50

Vol. 52: D. J. Simms, Lie Groups and Quantum Mechanics. IV, 90 pages. 1968. DM 8,–/$ 2.00

Vol. 53: J. Cerf, Sur les difféomorphismes de la sphère de dimension trois ($\Gamma_4 = 0$). XII, 133 pages. 1968. DM 12,–/$ 3.00

Vol. 54: G. Shimura, Automorphic Functions and Number Theory. VI, 69 pages. 1968. DM 8,–/$ 2.00

Vol. 55: D. Gromoll, W. Klingenberg und W. Meyer, Riemannsche Geometrie im Großen VI, 287 Seiten. 1968. DM 20,–/$ 5.00

Vol. 56: K. Floret und J. Wloka, Einführung in die Theorie der lokalkonvexen Räume. VIII, 194 Seiten. 1968. DM 16,–/$ 4.00

Vol. 57: F. Hirzebruch und K. H. Mayer, O(n)-Mannigfaltigkeiten, exotische Sphären und Singularitäten. IV, 132 Seiten. 1968. DM 10,80/$ 2.70

Vol. 58: Kuramochi Boundaries of Riemann Surfaces. IV, 102 pages. 1968. DM 9,60/$ 2.40

Vol. 59: K. Jänich, Differenzierbare G-Mannigfaltigkeiten. VI, 89 Seiten. 1968. DM 8,–/$ 2.00

Vol. 60: Seminar on Differential Equations and Dynamical Systems. Edited by G. S. Jones VI, 106 pages. 1968. DM 9,60/$ 2.40

Vol. 61: Reports of the Midwest Category Seminar II. IV, 91 pages. 1968. DM 9,60/$2.40